Frogs and Toads

Jules Howard

BLOOMSBURY WILDLIFE
LONDON • OXFORD • NEW YORK • NEW DELHI • SYDNEY

BLOOMSBURY WILDLIFE
Bloomsbury Publishing Plc
50 Bedford Square, London, WC1B 3DP, UK

BLOOMSBURY, BLOOMSBURY WILDLIFE and the Diana logo are trademarks of Bloomsbury Publishing Plc

First published in United Kingdom 2019

A catalogue record for this book is available from the British Library

Library of Congress Cataloguing-in-Publication data has been applied for

ISBN: PB: 978-1-4729-5581-4; ePub: 978-1-4729-5582-1; ePDF: 978-1-4729-5583-8

2 4 6 8 10 9 7 5 3 1

Design by Rod Teasdale
Printed and bound in India by Replika Press Pvt. Ltd.

Contents

Amphibian Apparel

You would be hard pushed to design from scratch an animal seemingly so weird as a frog or toad. Look more closely, however, and you will discover a story written in fossils that charts the rise of a group of animals that hit upon a shape and style almost perfect for dodging the interests of monstrous predators, yet that are themselves the stuff of nightmares for the invertebrate prey on which they thrive.

With their sticky tongues, bulbous eyes, madcap leaps and spectacular metamorphosis from tadpole into adulthood, amphibians sound like something dreamed up by a sugar-addled six-year-old. Shine a spotlight on these creatures, however, and a far more subtle and engaging reality emerges. There is real beauty in frogs and toads, but it is not a beauty you may ever have considered. And so, if I have one aim in this book, it is that you might come to view Britain's handful of frog and toad species in a new light. What we lack in the number of amphibian species we more than make up for in interesting stories of ecological endeavour and charismatic grace.

Above: Both frogs and toads have evolved in a world of large wetland predators. All species have large, sensitive eyes and are ever-watchful of threats.

Opposite: Mastering two domains, both water and land, means amphibians can take advantage of twice the ecological space in a single lifetime.

So, turn the pages that follow. Together, we will become more acquainted with the UK's frogs and toads and their unusual modes of life. And perhaps we will see it in our hearts to engage further with their conservation. For, as we will discover, the fate of frogs and toads across the world is far from secure, and these threats apply in the UK as much as anywhere else. But there is plenty to offer us hope, not least a vibrant conservation scene, citizen science at its best and a public that really cares. And amphibians have great persistence and survival in their blood. Perhaps, therefore, they have something to teach us.

Family relations

Where did frogs and toads spring from? And what, if any, is their family relationship? Frogs and toads both belong in the order of tailless amphibians called Anura, or the anurans, literally meaning 'without tail'. The first fossil anurans – amphibians with distinctive long leg bones, a three-pronged pelvis and a highly reduced tail – come from the Jurassic Period, dating back approximately 180 million years. Although it is tempting to consider those early creatures as primitive, not much further evolved than the early land-fish that first walked the Earth 370 million years ago, they were apparently proficient in all sorts of ways. Highly adaptable, quick to branch into new species and with long legs able to foil a host of predatory attacks, the earliest anurans hit upon a shape that would allow them to prosper throughout the Jurassic and into the Cretaceous Period while other species around them died out.

Above: The Natterjack Toad – a 'true toad' because it belongs to the bufonid branch of the frog family tree.

So, what is the difference between frogs and toads? Scientifically speaking, true toads are members of Bufonidae, or the bufonids, one family among 47 others that are placed within the order Anura. If this sounds curious to you, welcome to the occasionally confusing nature of taxonomy! Even more confusingly, there are amphibians we commonly call toads, like the European Common Midwife Toad (*Alytes obstetricians*), that are not true toads at all. In Britain, luckily, both our 'toad' species – the Common Toad (*Bufo bufo*) and Natterjack Toad (*Epidalea calamita*) – are true toads. For the purposes of this book, therefore, which focuses on UK species, the term 'toad' is used to refer to true toads only, while 'frog' is used to refer to all other anurans.

The Age of the Anurans

The question of when toads sprouted from the anuran family tree has become a hot topic of debate, but many consider today that they probably evolved in the Cretaceous Period (145–65 million years ago) in what is now South America or, intriguingly, a once ice-less Antarctica. In the millions of years that followed, as the land split apart and merged into the vast continents we know today, toads continued to diversify and formed new branching groups, each riding its given continent like a lifeboat. For this reason, many toad species today often occur over wide continental areas. The Common Toad, for instance, isn't just a British species; it has a distribution that stretches from the UK in the west, to Asia in the east and North Africa in the south. Not bad for a creature with little by way of speed, you might think.

In all, there are nearly 7,000 anuran species, although many new species undoubtedly remain out there for scientists to discover. Indeed, 60 per cent of new anuran discoveries have been made in the last 30 years. The extinction of the dinosaurs appears to have played an important part in the success of this charismatic order of life. By comparing the genetics of hundreds of living anuran species and matching them up with the fossil record to make a detailed family tree, it appears that a massive radiation of successful frog families emerged and branched out in the years that followed the great meteorite impact, 66 million years ago. After the (non-bird) dinosaurs ceased to be, it was anurans, undoubtedly, that prospered in a world rich in invertebrates until the larger mammalian and avian predators evolved. Indeed, anuran diversity today may be unmatched in geological history. This is as much the Age of Anurans as it is the Age of Mammals, one could argue.

Above: A fossilised frog from the now extinct genus *Palaeobatrachus*. These frogs evolved in the Cretaceous before their demise half a million years ago.

Wedded to water

Although nearly all frogs and toads are united in their need to visit water to lay eggs, members of this curious group of amphibians have adapted to life on land in all sorts of ways, sometimes breaking free of their wetland niche almost totally. The Turtle Frog (*Myobatrachus gouldii*) of Western Australia, for instance, spends most of its life underground, digging through dry sandy soils in search of termites, whose nests it breaks up using its sturdy front legs. On the whole, however, most frogs and toads are creatures of wetlands for at least a part of their life cycle.

Most anurans do little by way of parental care for their offspring. Many, if not most, species lay their eggs in clumps or strings and leave the tadpoles to fend for themselves once they have hatched. The odds of survival for most tadpoles are very low in most cases. For instance, there may be as many as 1,000 eggs in a blob of frogspawn laid by the Common Frog (*Rana temporaria*), of which perhaps only 50 tadpoles

Below: [illegible]

Above: The skin on the dorsal side of the Common Suriname Toad is an evolutionary marvel.

make it to metamorphosis and just two or three frogs reach reproductive age two or three years later. Every adult anuran alive today is a representative of fortune, therefore, a lucky one or two in a thousand.

Not all frogs and toads let the universe decide their fate in such a way; it seems that some anurans make their own luck. In habitats where predators abound, for instance, some frog and toad species have evolved to protect their offspring, guarding the eggs or even housing tadpoles within their bodies. One of the most curious examples of such parental care is the Common Suriname Toad (*Pipa pipa*), in which the male implants the fertilised eggs of the female in the soft skin of her uppermost (dorsal) surface. Within weeks, hundreds of tiny toadlets emerge en masse, essentially being 'birthed' one by one from the mother's back. Other anuran species that engage in parental care include the now extinct Australian gastric-brooding frogs, which kept their young safe inside the stomach, and the midwife toads of Europe and North Africa, whose males wrap egg strings around their legs, carrying them around in safety away from the pond.

Getting about

Above: [illegible] Wallace's Flying Frog makes [illegible] parachute [illegible]

A key characteristic of most anurans is their powerful back legs, which can be adapted for a variety of purposes, including jumping, digging and swimming. Aided by special tendons capable of releasing stored energy explosively, many anurans (frogs particularly) use their jumping as an effective means of escaping predators. Although many frog species vie for the title of longest jump, the American Bullfrog (*Lithobates catesbeianus*) is the one most ceremoniously put to the test in scientific literature. Laboratory studies have recorded a jump of 1.2m (4ft) in a single leap, though the species is likely to regularly exceed this in the wild. In North America, frogs were once regularly put to the test in public competitions which apparently saw individual frogs jumping past the 2m (6.5ft) mark. In the wild, smaller anurans may regularly exceed even this impressive figure.

The evolution of webbed toes and loose skin flaps on their sides has given many treetop anurans the ability to glide. Wallace's Flying Frog (*Rhacophorus nigropalmatus*), for instance, is capable of leaping more than 15m (50ft) from tree to tree, putting almost all other frogs to shame.

Above: The American Bullfrog [illegible]

But not all frogs are built for jumping or gliding. Some anurans, including many toads, have shortened legs, and individuals mostly crawl rather than leap. Many toads also use their back legs for digging small refuges in soil, sand and leaf litter.

Nearly all anurans are impressive swimmers, their long legs co-opted for moving through lakes, ponds and slow-flowing rivers with ease. Indeed, it is no surprise that some frogs and toads have evolved to conquer this aquatic domain more fully. These aquatic anurans include clawed frogs, representatives of the African genus *Xenopus*, which have flattened egg-shaped bodies with pronounced sensory organs not unlike those of some fish. Occasionally, clawed frogs come out of their aquatic realm, slipping and sliding their bodies through wet grass (rather like freshwater eels) in search of new ponds and lakes. Another anuran, the Crab-eating Frog (*Fejervarya cancrivora*) has even evolved to tolerate salt water for short periods, which it does by regulating the production of urea. Its tadpoles can withstand 3.9 per cent salinity, which is higher than in seawater.

Below: Frogs of the *Xenopus* genus have evolved a way of life that is almost fish-like.

Acoustic accomplishments

Nearly all anurans are highly vocal, with most species having a distinctive call unique to its species and made by squeezing air through the larynx in the throat. Many species amplify these laryngeal sounds through the use of vocal sacs, which act a little like a simple echo chamber. In some species, like the Common Frog, these vocal sacs can be inflated under the mouth. In others, such as the Pool Frog (*Pelophylax lessonae*), the sacs form from pouches at each corner of the mouth. Anuran calls are normally emitted by males during defined breeding periods, which in temperate regions occur during warm spring nights. Where multiple males converge on a breeding site, the intense chorus can sometimes be heard from many hundreds of metres away. But what sounds to us like a wall of sound is to female anurans a ticker-tape news stream relaying the sexual status of local males. On the whole, the females use these calls to judge male quality; often (but not always) they are interested in males with the deepest calls – a signal of size and quality that is hard to fake, though many try.

Below: [illegible]

Above: Common Frogs are at their most noisy when they gather together. Their call sounds more like a pained groan than the frogs we hear in Hollywood films.

Male and female anurans process these calls through some fairly impressive auditory anatomy not totally unlike our own. Although it is arguably more primitive than the mammalian ear, they possess an eardrum, called the tympanic membrane. This can be seen clearly in some frogs and toads as a distinct ring of skin behind and slightly below the eyes. Stretched like a drum, this skin membrane detects sounds vibrations, which are then processed in the brain as acoustic signals – sounds. Although anurans can hear the sound of approaching predators, predictably, much of their auditory capability is 'tuned' in to the frequency of their species' calls.

Not all anuran calls are about attracting members of the opposite sex. If approached and advanced upon sexually by another male, male Common Toads emit a special chirp (a release call) that signals to the rival that the interaction may be non-fruitful, so to speak. And some anurans can use their calls to scare away predators. There are many cases of Common Frogs emitting a shrill scream upon being grabbed by a predator (indeed, I have heard this myself – at times, it sounds eerily close to the cry of a human baby). It is likely that the noise may surprise predators, including rats and cats, and, some scientists suggest, may even serve to attract other predators that might then scare away the attacker.

Frogs and toads in the UK

Of the UK's native anurans, we have two resident families that each comprise two species. There are two species of true toads or bufonids, the Common Toad and the Natterjack Toad, and two species of true frogs or ranids (members of the family Ranidae), the Common Frog and the Pool Frog.

Telling true toads apart from frogs can be problematic, but there are some general features that can help would-be observers get a feel for which is which, in the UK at least:

Frogs	Toads
Skin is often smooth and moist	Skin is warty
Spawn is laid in a ball or clump	Spawn is laid in paired or single strings
Normally found within 300m (330yd) of water	Normally found within 1km (0.6 miles) of water
Jump frequently, particularly when alarmed	May run or walk rather than jump
Tend to remain near water or within long grass in summer	During hot spells toads may hide underneath logs and stones
Lack poison glands	Possess two poison glands behind the eyes
The Common Frog has a yellow iris and a circular pupil. In the Pool Frog (and other green frogs), the pupil is slightly flattened horizontally	Common Toads and Natterjack Toads have a horizontal slit-like pupil. In the Common Toad, the iris is orange, and in the Natterjack it is blue-green

Above: The Common Frog has notably stripy legs.

Above: A Common Toad with its charismatic fiery eyes.

Key features

As adults, frogs and toads are consummate predators. Every adult anuran you have come across attained its size by hunting invertebrates, especially insects, earthworms, slugs, snails and spiders. In their own special way, they are built for the hunt. The body plan of anurans consists of slightly protruding eyes on the top of the head, a sticky tongue that is often capable of being launched at passing prey, and a brain quick to process the movements of potential prey. But life as an anuran is tough. For starters, frogs and toads must dodge the interests of predators higher up the food chain, including mammals, birds and snakes. To counter this, many species have evolved a suite of adaptations to deter their interests.

All the better to see you

The bulbous eyes of frogs and toads, often comically exaggerated by cartoonists, are a key piece of sensory anatomy and are large for a reason. By bulging from the skull, the eyeballs provide binocular vision when looking forwards, while offering a total visual field of almost 360 degrees to spot approaching predators. In this

Below: The bulbous eyes of the Common Frog can be pulled into the skull to help push food down the throat.

way, they are like photon-gathering satellite dishes. Frog and toad eyes come in a variety of iris shapes and pupil shapes, which can be used to identify species. Common Toads, for instance, have orange irises and horizontal slit-like pupils, while Common Frogs have a circular pupil within a yellow iris.

Anuran eyes are particularly sensitive to movement and their attentions are quickly drawn towards worm-shaped objects in their field of view, particularly if these items wriggle or crawl. Watching what happens next is enthralling. Upon sensing movement, the frog or toad will turn its entire head toward the object of interest, rather like a security camera focusing in on a suspect. At this point, the frog or toad will often pause in an apparent trance for a number of seconds, watching its prey with rapt intensity. Sometimes, whole minutes can pass with the animal in this state, which can be rather frustrating for human observers. Then suddenly... SNAP!... the prey is gone. The reason for this occasionally long pause is interesting. It may be that the amphibian is getting its calculations straight – its vision of the prey is momentarily lost when it strikes with its tongue, so it must line everything up perfectly before committing.

Below:

With the prey captured, now comes the *coup de grâce*. The frog or toad pulls its eyeballs deep into its sockets while swallowing, using them to plunge the prey downwards into the stomach. Although it may seem a little disturbing to us for a creature to use its eyeballs for such a purpose, this is, by and large, the amphibian way. Within seconds, the eyes pop back into place on the top of the skull, and the frog or toad gives its mouth and eyes a little wipe. It is now armed and ready for the next prey item that strays too close.

Above: If spooked, the Common Frog quickly makes for the cover of water. Often, a single leap will do.

Great leap forward

Undoubtedly, a key part of the initial evolutionary success of frogs and toads is down to their long hind legs. This unusual limb anatomy has been exploited by various frog species in a number of ways, particularly for swimming and (perhaps most importantly) as a means of leaping. In all modern-day anurans, many of the leg and foot bones have fused into a single, stronger bony structure that more effectively allows its owner to 'catapult' the body forward. These fused leg bones also absorb impact upon landing, limiting the potential for broken bones and wear-and-tear fractures. Anurans also have an unusual fused tailbone structure, the urostyle, which directs force from the legs towards the body during a leap. Predictably, anurans are endowed with impressively powerful leg muscles – at any one time, these muscles may account for as much as 17 per cent of the total body mass.

Slow-motion photography has helped scientists learn more about how these impressive jumps are achieved.

As in grasshoppers and locusts, much of the springing motion comes from the sudden release of energy stored in stretchy tendons attached to the muscles. When many anurans prepare to jump, they stretch these tendons out before letting them recoil, assisting the muscles in powering the jump forwards.

Not all anurans depend on such leaps though. Toads, particularly, appear to have lost this specialism over time. Although they can jump a little, their efforts are pretty pathetic in comparison to frogs. Instead, they use their back legs for digging. By churning their legs back and forth, they can submerge their bodies into soft ground – useful for hiding from predators or seeking shelter from the hottest and coldest spells of weather.

Above: Water evaporates easily through the skin of frogs. This is one reason they prefer damp places.

Under the skin

When it comes to digestion, the internal anatomy of frogs and toads is not a great step forward in sophistication from the terrestrial land-fish from which they evolved. They possess many of the same familiar structures as fish, including an oesophagus, stomach and intestines. Digestion is achieved, as in most fish, using pancreatic juices and bile produced by the liver and stored in a gallbladder. Broadly speaking, all land-living bony creatures have that in common, but there are things that set amphibians apart from their land-living cousins, the reptiles and mammals. Like fish, many amphibians lack the ability to manage water loss effectively in very dry environments. Their skin is permeable to water, which is therefore lost directly from it by evaporation. But amphibian skin can do something ours cannot: on the lower (ventral) surface it can absorb water from puddles and damp soil, which is then stored in a bladder. In desert species, this stored water can account for one-third of an amphibian's total body weight.

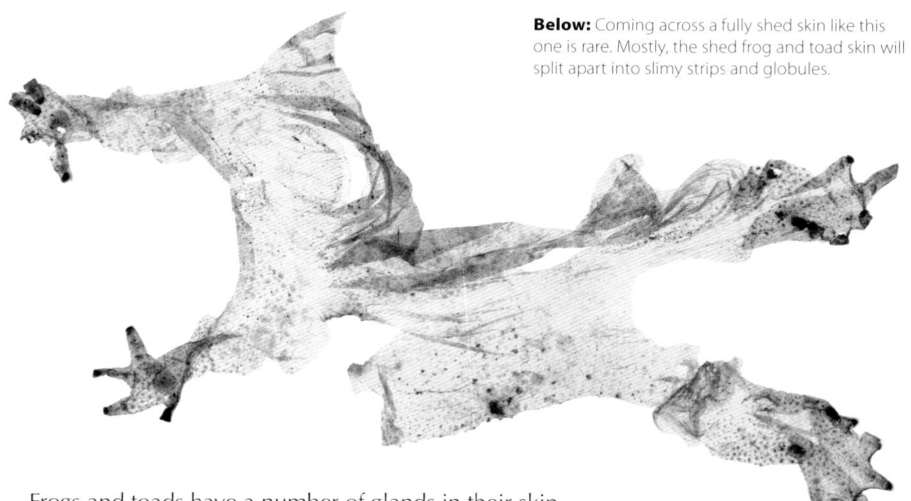

Below: Coming across a fully shed skin like this one is rare. Mostly, the shed frog and toad skin will split apart into slimy strips and globules.

Frogs and toads have a number of glands in their skin that ooze sticky secretions, many of which keep the skin moist. In toads, these secretions have been co-opted to become poison glands, producing toxins that make the skin distasteful to predators, which may then prefer to spit them out than attempt to swallow them. These glands also produce chemicals that protect against disease (see page 46).

Both frogs and toads can be masters of camouflage and, when it comes to their pigmentation, there are often impressive variations between individuals of a species. Both the Common Frog and Common Toad are highly variable, and orange, black, brick-red, lime-green and even pink specimens turn up from time to time. Like reptiles, all amphibians shed their skins. Every few weeks or months, the upper side of the frog or toad will split across the middle and peel downwards, exposing a new layer of skin underneath. Anurans sometimes eat this discarded skin, a habit that could be aimed at removing local odours that would otherwise attract predators, or it may serve to recycle nutrients – or it could be that the skin simply tastes nice.

Below: A Common Frog skeleton complete with fused tailbone structure (the urostyle), which directs mechanical force towards the legs.

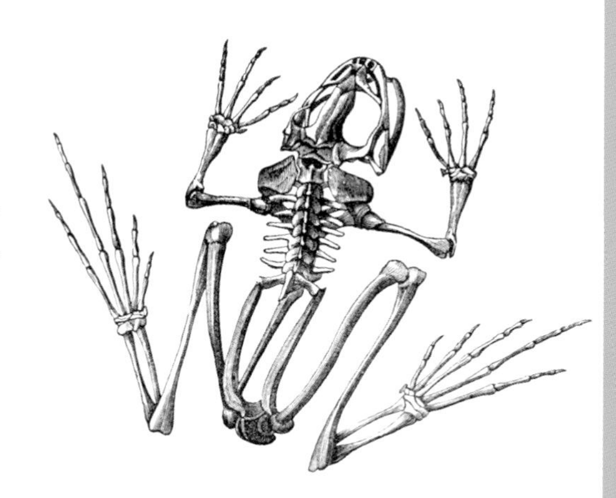

Meet the Residents

The UK is home to two native species of true frogs (ranids) and two species of true toads (bufonids). Each has its own unique habitat preferences, ranging from lakes and reservoirs to heathland ponds and temporary pools. Some, like the Common Frog, even do rather well in ornamental garden ponds and fountains. This chapter puts the spotlight on each species, describing their key features and preferred habitats, and exploring their unique behaviours and charisma.

Common Frog

This is the UK's most widespread and easy-to-observe amphibian. Quick to colonise new ponds and explore new areas, the Common Frog (*Rana temporaria*) can take up home in even the smallest of ponds. Indeed, this is a species that has moved increasingly into urban areas, riding on the back of the modern human trend for garden water features and ornamental ponds.

Opposite: A male and female frog readying themselves for egg-laying. Note the blue tinge to the male's white throat.

Below: The nuptial pad of a male Common Frog.

Adult identification

The Common Frog is a relatively smooth-skinned amphibian with a rounded snout and conspicuous black 'goggle' patches behind the eyes (think Zorro). Adults measure 6–9cm, from nose to tail. Individuals normally have varied black mottling across their back, and their legs are often marked with stripy bands. During the breeding season, the males usually have a blue tinge to their white throat and they also possess enlarged black 'thumbs' – the so-called nuptial pads – which they use to grasp females. Females are harder to identify upon first observation, although they are often larger (particularly before breeding) and their flanks may be covered with a mottled red or orange pattern.

Common Frog calls differ from those of Common Toads. Their breeding call is not quite a croak, but more of a creak or groan. To me, it sounds like boots walking through thick, crushed snow, although others liken it to a purr.

Above: In spring, males also have a blue tinge to their white throat.

Above: [illegible]

Habitats

Common Frogs are normally found within 300m (330yd) of still or slow-running fresh water, be it lakes, streams or ponds. Most will forage in damp habitats such as woodlands, moorlands, wet grasslands, parks and, of course, gardens. On sunny and dry days, however, wandering Common Frogs tend to retreat to water or long grass. Spawning sites include the margins of lakes, shallow or seasonal ponds, and even large puddles or ditches. Some individuals have even been known to spawn in water-filled tractor tracks or upturned dustbin lids. On the whole, though, Common Frogs prefer ponds without fish. In the UK, they normally breed slightly earlier than Common Toads. In the north and central regions of the British Isles, spawning tends to occur after

Right: [illegible] wet grass is a [illegible] habitat for frogs in late summer.

a string of mild, wet nights in March. In the south and west, it can be as early as January or February. In East Anglia or the far north of Scotland, spawning can occur as late as April.

UK distribution

During the latter half of the 20th century, agricultural expansion and the loss of traditional farmland ponds spelt doom for many amphibian populations across the UK, but undoubtedly the Common Frog has benefited from our modern propensity for urban fresh waters, including ornamental ponds. Today, Common Frogs occupy many waterbodies throughout Britain and Ireland, in places as remote as the Hebrides, Guernsey and the Isle of Man. In Scotland, they are even able to colonise ponds at altitudes above 1,000m (3,300ft). Interestingly, Common Frogs do not occur in Jersey, where the Agile Frog (*Rana dalmatina*) is found instead. This long-legged, athletic frog is a close cousin of the mainland species.

Pond pioneer

The Common Frog has a host of parochial names that turn up from time to time. Locals in the north of England used to call it a frosh or a fresher, while people living in the south-west nicknamed it a glouton, gwelsken, paget-poo (Cornwall) or pan (Somerset). Froglets (young frogs) also had some rather delightful local names, including frosks, laid-licks, charlies and, in Scotland, puddocks.

Below: Frogs like open and sunny ponds that allow for the growth of algae, upon which their tadpoles depend.

Above: [illegible]

Being relatively widespread and unfussy when it comes to breeding ponds, the Common Frog is the amphibian most of us encounter first as children. Many herpetologists (as those who study amphibians and reptiles are called) consider them formative components of their later careers. For me, the sight and consistency of frogspawn in a school pond takes me right back. I can remember the alienness of the substance, and the awe that it inspired within me, as if it were yesterday. Because of their dogged determination to find new ponds, Common Frogs have impressed their strange ways into the lives of many pond-owning urbanites. I should know; I used to run a frog helpline. Every spring, the helpline would receive hundreds of calls from gardeners confused or delighted by the writhing and wrestling sexual antics of their amphibian residents. Common Frogs really do have one-track minds at this time of year, so this makes them very accessible animals to get up close to and study. Indeed, at one point Common Frogs were the go-to animal for scientific study – so much so, that wild populations in some areas were ravaged by young children catching them and selling them on to meet demand for the trade.

Common Frogs have so taken to garden ponds that it is interesting to consider what they might have used in a world before these features were available. In those heady days, ancestral breeding grounds would have included oxbow lakes, tarns and pingo ponds or kettle lakes – water-filled depressions left over after giant glacial ice masses melted beneath the soil – as well as lakes and slow-flowing streams. Many Common Frogs will also probably have used the pockets of water that occur where large trees fall over, pulling up the soil and exposing the roots in a great cleft. Many more will have benefited from a close association with that master pond builder, the Eurasian Beaver (*Castor fiber*), which became extinct in the UK in the 16th century. In fact, the current reintroduction of Eurasian Beavers to some parts of their former range is good news for Common Frogs and other amphibians.

Common Frogs are nothing if not tenacious. It is incredible to consider that every individual in mainland Britain is the descendant of a few pioneering frogs that crossed the land bridge from Europe after the last Ice Age 10,000 years ago. Ireland's frogs have a different story to tell. In 2009, DNA studies confirmed that they had somehow found an ice-free refuge to see out the last of the ice ages. It seems that the attachment of Irish Common Frogs to their home goes far deeper.

Above: Some common frogs have a black mottled pattern (like spots of ink) unique to each individual – a bit like a human fingerprint.

Common Frogs are creatures of boom and bust. What starts off as a few blobs of frogspawn in some ponds can turn into an apparent glut of spawn a few years later as populations grow seemingly exponentially. These sudden fluctuations in local frog populations can worry many pond owners, who may have concerns about water quality or the welfare of their fish, but then numbers invariably fall. As competition kicks in, survivorship decreases – there is less for each individual and the population plummets. This is the way for many amphibians, and probably the reason that stories of so-called 'plagues' of frogs are pervasive across human cultures (see page 90). In some years, for some frogs, the world is a fruitful place for their generations. But local resources are never limitless – a lesson we humans might do well to respect more often.

Pool Frog

Above:

Lost to extinction in Britain in the 1990s, the Pool Frog (*Pelophylax lessonae*) has now returned, courtesy of a number of carefully planned reintroductions to a network of fenland pools in Norfolk (see below). Today, these new populations are faring well. Loud, vivacious and more aquatic than other British frogs, Pool Frogs have a breeding season that extends over a longer period than in other amphibians, sometimes with a second spawning in early summer.

Adult identification

The Pool Frog measures 6cm (2.4in) from nose to tail. This species often has a distinct yellow or green stripe running along its back, and its snout is notably more pointy than that of the Common Frog. Unlike the Common Frog, the bulging white vocal sacs come from the sides of its mouth rather than from underneath. Its breeding calls consist of a series of loud croaks and purrs, which some people liken to the cries of a duck.

Habitats and UK distribution

The last known haunts of British Pool Frogs before they disappeared were pingo ponds in parts of Norfolk.

However, based on the evidence of partially fossilised finds, it is likely that native Pool Frogs once occupied ponds across much of East Anglia. These were the 'true' Pool Frog natives, and shouldn't be confused with the 'false natives' – populations of Pool Frogs that were artificially established across England in the 20th century, and that have subsequently hybridised with other closely related species from mainland Europe, namely the Edible Frog (*Pelophylax kl. esculentus*) and the Marsh Frog (*Pelophylax ridibundus*) (see page 37 for more on this).

Above: Pingo ponds are in-filled impressions in the ground made when solitary chunks of ice melted after the last Ice Age 10,000 years ago.

The reintroduction of true Pool Frogs to eastern England has been incredibly encouraging. After the successful translocation of individuals from the first breeding colony established in 2005, a new breeding population is now taking root thanks to the work of the Norfolk Wildlife Trust and Amphibian and Reptile Conservation (ARC), in partnership with a host of organisations including Natural England and the Forestry Commission. Through their careful coordination a UK species almost lost forever is now back where it belongs.

Below: Like other so-called 'green frogs', the Pool Frog likes to bask in the sunshine at various times of the day.

A conservation success

For about a century, the status of Pool Frogs in Britain was one of the most fiercely debated topics among herpetologists. Was the dwindling Norfolk population truly a British species or was it simply a remnant population of introduced frogs brought over from Europe? At the end of the 20th century, DNA and fossil evidence appeared definitive, proving that they were indeed native. Sadly, by this time it was too late: the British Pool Frogs were down to one last individual, the ironically named Lucky. Caught from a pond in Thetford in 1993, Lucky was an 'endling' – the last of her species in Britain. She died in captivity in 1999, an event that seemingly signalled the end of the whole sorry saga. The Norfolk Pool Frogs had been native, but they were now, sadly, extinct.

Above: Pool Frogs produce a loud call through a pair of vocal sacs (shown) at the corners of the mouth.

Thankfully, however, this was not the end of the story. By assessing the genetic make-up of European Pool Frogs, conservationists and researchers in the UK homed in on Swedish populations, which proved a close genetic match for Lucky. Essentially, our frogs were like their frogs. Between 2005 and 2008, using Pool Frogs collected from the wild in Sweden, a team led by ARC (then called the Herpetological Conservation Trust) re-established a British population. Since this introduction, the frogs have done rather well, and from this first population a second population has been established. It is likely that, in time, more will follow. For many frog lovers, this has been a wonderful and inspiring project, although some critics of the project do remain. For their part, the Pool Frogs continue to do what they do best: making more Pool Frogs. And long may they do so.

I have only seen Pool Frogs once in the UK, and it was a genuine privilege to be among them. Being used to Common Frogs, I was struck by their slightly smaller size and seemingly boundless energy. Unlike their common cousins, which tend to aggregate in a single part of the pond, it felt like the Pool Frogs were everywhere, warring for position on floating vegetation and calling from the sun-drenched banks. They were basking, too, rather more like lizards than frogs as we know them. The sex lives of frogs and toads generally play out in early spring, sometimes while there is still frost on the ground. Yet here were Pool Frogs calling alongside summer migrants like the Cuckoo (*Cuculus canorus*) and Hobby (*Falco subbuteo*). I suspect that most members of the general public consider one frog to be very much like another. But Pool Frogs are different from Common Frogs in many ways, so it is a great thing to be able to consider them true British residents, resurrected.

Above: If the Pool Frog recolonises its former haunts, future generations may get to see the fruits of our conservation success.

If there is any sadness in the tale of the Pool Frog it is this: for the moment at least, very few people in Britain stand a chance of seeing this treasured amphibian, so isolated are its specially managed (and protected) East Anglian habitats. That said, there are plans to connect existing Pool Frog sites, encouraging the amphibians to spread across their former haunts. Although most of us may never come face to face with a Pool Frog, our children, or our children's children, hopefully will. And when they do, they will, I hope, consider momentarily the conservationists who made it all happen.

Common Toad

Above: [illegible]

Few amphibians have captured the public attention quite like the Common Toad (*Bufo bufo*). With its fiery eyes, poisonous skin and apparent nonplussed attitude towards humans, it is no surprise that this is a species about which folklore abounds. Look more closely at Common Toads, however, and you will see a once widespread species on the slide, increasingly limited in its habitats and behaviours by human activity.

Adult identification

Common Toads (most individuals measure up to 8cm or 3.1in from nose to tail) can be identified fairly easily by looking for a trio of key features. First, they have dry, warty skin with two large bumps behind the eyes – the poison (parotoid) glands. Second, the eyes are normally orange or even copper-coloured, with a horizontal black pupil. And third, Common Toads prefer to walk rather than jump, and are more likely to sit still when they come face to face with a person rather than leap energetically away in a frog-like manner. Like Common Frogs, however, individual Common Toads display impressive variations in their colour and markings, some individuals being grey, grey-brown, black, yellow or even brick red. Their undersides are often white and mottled. In spring, the call of the male sounds a little like a squeaky dog toy.

Habitats

With their notably thicker skin, Common Toads can roam much further from fresh water than most other amphibians. Their hunting grounds include heathlands, gardens, forests and moorlands, and they will often sit quietly underneath logs and stones, ambushing any invertebrate prey that may accidentally stray too close. Compared to other amphibians, Common Toads have marathon-like endurance: some will regularly make breeding migrations of more than 1km (0.6 miles) back to ponds, covering the distance in only a few nights. Common Toad breeding ponds include large ponds, lakes and reservoirs. They are less fussy than Common Frogs about the presence of fish, probably because their tadpoles, like the adults, are poisonous. Most Common Toads will stay loyal to breeding sites, returning to mate in the ponds in which they themselves developed.

Below: Unlike the Common Frog, the Common Toad has a distinctive horizontal pupil across each eye.

UK distribution

Common Toads are found throughout mainland Britain but are notably absent from Ireland. Though once also considered residents of the Channel Islands, scientists now consider Jersey's toads to be a closely related species, the Spiny Toad (*Bufo spinosus*).

Dogged and determined

Like the Common Frog, the Common Toad has amassed a rich amount of folklore, much of which references medieval perceptions of the creature as one shrouded in mystery, magic and foreboding fortune (see page 92). Some local names for Common Toads in Britain include gangril (northern England), hornywink (Worcestershire) and slug (Cornwall), along with the fabulous paddock, puddock or paddock-rude, all of which are derivatives of the Norse word for toad, *padda*.

Compared to Common Frogs, Common Toads tend to be shy creatures, rarely exposing themselves to humans except during the breeding season. When they do turn up – for example, when objects they are hiding under, such as logs, are upturned by inquisitive naturalists – they tend to stand firm, eyeballing intruders with a cold, hard stare before quietly moving onwards to a nearby dark spot. Common Toads rarely hop, preferring instead to move in a rugged crawling manner.

For many years, it was thought that Common Toads spent winter lying dormant beneath the roots of trees and in other such holes, but there is good evidence that many simply squirm their body into the soil, burying themselves downwards with their back legs to provide protection from frost and ice. Many gardeners report discovering toads while digging flowerbeds in winter months. In this condition, the amphibians' skin often appears darker – an adaptation, one assumes, that helps it retain heat.

In spring, the mass migration of Common Toads to ancestral breeding ponds is one of nature's most impressive spectacles. They have a sort of doggedness at this time of year, as if pulled towards the ponds by an invisible string. Common Toads often make these movements approximately one or two weeks after

Above: Cars and lorries can be a serious threat to many toads during their breeding migrations in spring.

Common Frogs, although the breeding of the two species can overlap. Often, toad migrations begin in the south-west in February and activity spreads northwards throughout central and northern regions in March. In Scotland, the north of England and East Anglia, spawning can take place early in April. At this time of year, as their migrations begin, one sees a true picture of local toad populations. In some places, their numbers can really impress. Thousands of toads converge on some sites in the UK, often within a few days of each other. During balmy spring nights, these ponds come alive with their writhing and wrestling bodies, and the air is filled with their chirping calls. Large males fight with one another, while smaller males watch from the sidelines, looking to steal a moment with an unattached mate. Females splash and spiral through the water, their bodies intertwined with sometimes five or more males. And then, a few days later, all goes quiet and the toads are gone. The resulting tadpoles face an uncertain fate in the pond, but at least they are armed with the same weapons as their parents: like the adults, Common Toad tadpoles are poisonous, and many fish prefer to avoid them.

Natterjack Toad

Right: [illegible] note the [illegible] stripe along [illegible]

Natterjack Toads (*Epidalea calamita*) are specialists of dry heathlands and sand dunes, where they breed in shallow pools and dune slacks. Once threatened with extinction in Britain and Ireland, and reduced to isolated populations, the species is now recovering thanks to concerted actions by conservationists in recent decades. Today, about 60 Natterjack sites exist around the UK, all of them protected by law to prevent further declines.

Adult identification

The Natterjack Toad (6–8cm or 2.4–3.1in from nose to tail) is often slightly more diminutive than the Common Toad in size. It has complex marble-like patterns overlying a light brown background that blends to white on its flanks. The yellow dorsal stripe that runs from the head to the base of the spine provides a handy diagnostic tool for would-be amphibian spotters. As in Common Toads, the skin of Natterjacks is also warty and two distinctive parotoid glands are visible behind the eyes. Natterjack breeding calls are loud and, when performed in unison, can carry for many hundreds of metres. When multiple males sing together, a chorus forms – a distinctive chirring that sounds like a grasshopper song but many, many times louder.

Below: The horizontal pupil and [illegible] patterns of the iris give [illegible] a marble [illegible]

Above: The Natterjack Toad calls by inflating a vocal sac under its chin. Its call can be heard up to a mile away.

Habitats

National strongholds for Natterjacks include the coastal dunes of Merseyside and Cumbria, and the saltmarshes of the Solway Firth, although scattered populations exist elsewhere, particularly in Lincolnshire and Norfolk. Reintroductions to former Natterjack haunts have also been trialled successfully in north Wales and Kent. Many Natterjack breeding grounds are on public land, but their protected conservation status means that disturbance is strictly prohibited.

UK distribution

The Natterjack Toad occurs in discrete populations across England, Wales, Scotland and Ireland, including at several dune sites in County Kerry.

A nightingale almost lost

The name Natterjack comes from the ratchet-like call made by males, a noise long associated with the species. So intense is this chorus that, for many years, Natterjacks have gone by a number of alternative local names, including the Birkdale nightingale in Lancashire, the Thursley thrush in Surrey, and, on Merseyside, the Bootle organ.

Other local names for Natterjacks refer to their fast-paced walking style – hence the term running toad, which

Above:

was once used quite commonly. It is easy to see why such a name would stick when observing the movements of Natterjack toadlets, which run quickly on rigid legs and keep their bodies held high, leaving a trail of distinctive, tiny mouse-like footprints behind them. Undoubtedly, this strange walking style is an adaptation to keep cool on hot sandy soil. This, and their distinctive jewel-like blue-green eyes, gives Natterjacks a special kind of charisma.

Although Natterjacks are rare and highly protected, guided walks at Natterjack sites in the UK in April and May offer visitors the opportunity to see (and hear) the toads in their breeding splendour. These events can be lots of fun, and also provide opportunities to spot other rare creatures, including, in some places, Sand Lizards (*Lacerta agilis*).

The scale of the 20th-century decline of Natterjack Toads is nothing short of jaw-dropping and speaks volumes about the dramatic loss of the dry, sandy heathlands with which they associate. It is estimated that only one-sixth of the heathland in the UK that existed 200 years ago remains today, with the surviving patches facing a suite of pressures, including air pollution, recreational disturbance and human-caused heathland fires. Thankfully, all of our Natterjack populations are now very well managed and their survival in the wild – albeit in restricted populations – is secure from further losses. Natterjacks are likely to be with us a good while yet.

Amphibian invaders

Although it is now illegal in the UK to release non-native species and has been for decades, small residual populations of non-native frogs and toads still live in the wild. These are often the progeny of once captive garden collections that have since run amok.

European imports

The most common of the UK's introduced anurans are the green frogs or water frogs, which are a genetic mishmash of closely related frog species that thrive in mainland Europe and were introduced to Britain in the early 20th century. In addition to our native Pool Frog, the green frog 'complex' includes the Edible Frog and the Marsh Frog. All of these closely related frogs are noisy and like to remain at the water's edge. They also have a pair of vocal sacs that inflate behind the corner of the lips rather than underneath the chin, as in Common Frogs. Small populations established between 1903 and 1960 in Surrey have spread along riverbanks into East Sussex, where they thrive today. Other populations exist in Hampshire, Somerset, Yorkshire and – as if to signpost the problem of non-native species reaching British shores – near Heathrow Airport. The large population that now exists in Romney Marsh originated from just 12 frogs introduced from Hungary in 1935. Green frogs are also found at the London Wetland Centre, a popular tourist spot.

Of the exotic toads found in the British Isles, perhaps the tiniest, yet enduring populations are those of the

Below: Green frogs often bask on the sunny side of ponds. If startled, they quickly dart into the water.

Above: [illegible]

Common Midwife Toad, which, it is thought, snuck into Bedfordshire in the early 20th century through European imports of nursery produce. This is quite a story. Rehomed in the garden of a local family, the toad's numbers grew and further 'rehomings' took place, courtesy of the family's son, who took some of the amphibians to his own home in Woodsetts, near Worksop, an area in which they still thrive today. From these early introductions, the species has spread, particularly across north Bedford. The presence of these tiny toads – which are about half the size of the Common Toad – is often given away by their high-pitched calls in late spring, which sound rather like the noise a smoke alarm makes when the batteries need replacing: a high-pitched *BIP!* The common name refers to the male toad's unusual parenting behaviour. Once the female's eggs are fertilised, the male wraps them around his hind legs and pulls them from the water, walking them around safely on land to give them time to develop away from the watchful eyes of pond predators.

From further afield

In the past, other populations of non-native frogs and toads have temporarily made the UK their home, before dwindling towards extinction or being removed (often at great cost) by authorities. They include the African Clawed Frog (*Xenopus laevis*), which was once, and may still be, resident in some ponds in south Wales, and the European Tree Frog (*Hyla arborea*), once known to have lived in isolated parts of London, Devon and the Isle of Wight. A mysterious colony of European Tree Frogs is also known to have existed in the New Forest until at least 1988.

When it comes to invasive non-natives, the UK's *bête noire* is the American Bullfrog. These large frogs are capable of laying up to 60,000 eggs in a single spawning event, and adults are even known to eat smaller amphibians, including native species. Once a commonly traded pet, the species is now banned from importation to the UK due to its propensity for establishing wild colonies when thoughtlessly released into the wild.

There may well be other introduced populations of frogs and toads about which scientists know very little. If seen in the wild, non-native species can be reported through the National Amphibian and Reptile Recording Scheme's *Alien Encounters* website (alienencounters.narrs.org.uk).

Below: The American Bullfrog is a highly invasive species that has successfully bred in the wild in the UK.

Life on Land

Over millions of years, frogs and toads have adapted to life on land. These highly mobile invertebrate-devouring predators are armed with a range of special features to allow them to spot and catch prey. But feeding is only half the battle. The amphibians themselves are threatened by larger predators, from which they must flee. They manage this in a host of unusual and celebrated ways, as we discover here.

What do frogs and toads eat?

In the UK, the Common Frog has long been heralded as the 'gardener's friend' because of its propensity for feasting upon snails and slugs, but this is far from all it eats. Studies of the species' diet suggest that perhaps only 25 per cent of its prey comprises these terrestrial molluscs. Common Frogs also take woodlice, ants and, occasionally, creatures on the wing, like butterflies and moths. Being creatures of drier habitats, Common Toads prey more on beetles and ants, the latter making up perhaps 40 per cent of their diet. This minimal overlap between the diets of the two species means that they can, and often do, live quite happily in close proximity. Natterjacks also eat ants and beetles, but in addition, fly larvae appear to be particularly important, perhaps accounting for one-third of all prey consumed.

Before snapping at potential prey, frogs and toads must carefully ascertain whether an object

Opposite: The Common Toad's eyes are highly sensitive to the movements of a variety of terrestrial invertebrates.

Below: Snails, slugs, wood louse and ants are a few of the food items on the menu of the UK's frogs and toads.

Above:

that has caught their attention is a source of prey or a potential source of danger. They appear to do this by measuring the visual angle of moving objects, taking into consideration the distance between themselves and the object of interest. Larger objects elicit a defensive behaviour such as jumping away or, in Common Toads, the crouching 'predator-avoidance' stance. Smaller objects – particularly horizontal worm-like objects – elicit the prey response.

As well as being adept at spotting and swallowing passing prey, frogs and toads display impressive fasting skills. It's likely that most of our anurans stop feeding in late autumn when invertebrate populations dip, and then enter a dormant phase from which they may only occasionally rouse themselves. Likewise, both frogs and toads can see out dry, hot spells buried in mud or under logs and stones. Called aestivation, this period of dormancy – primitive in anurans – is an adaptation more commonly associated with some crocodiles and salamanders living in warmer, occasionally very parched, environments.

What eats frogs and toads?

A murder of crows

In 2005, Germany became the centre of an international wildlife mystery. Next to a pond in a public park one morning, passers-by noticed hundreds of dead Common Toads that looked as if they had, to all intents and purposes, exploded. From a hole in their undersides, their entrails littered the site, providing onlookers with a nightmarish vision akin to something from a science-fiction film. Not surprisingly, this apparent crime scene caught the attention of the global media, which for a period of five or six weeks discussed possible causes, including pollution and mystery viruses. One crackpot theory even suggested that it was a strange form of ritual witchcraft, undertaken not by witches, but by the toads themselves. The truth, when it was discovered, was almost as strange: crows were the culprits.

A handful of crows had learnt a neat trick, a behaviour that spread throughout the local crow population in a phenomenon known as horizontal learning. Some of these wily and inquisitive birds had realised that, by flipping Common Toads onto their backs, they could bypass their poisonous defences. Once the toads were immobilised, the crows pecked through the softer skin

Below: The corvid family includes a range of predators able to overcome the toad's poisonous defences.

Above:

on their underside and pulled out the internal organs (the liver particularly), leaving the toads looking like they had exploded. Poison glands can afford only so much protection, it would seem. But crows are just one example of a myriad of predators that can make meals of toads and frogs.

Being primary predators – middle rungs on the energetic ladder of wetland ecosystems – frogs and toads themselves are a source of prey for a host of predators. In some gardens, amphibian predators include owls, Foxes (*Vulpes vulpes*), Badgers (*Meles meles*), Grey Herons (*Ardea cinerea*) and even Hedgehogs (*Erinaceous europaeus*). Domestic cats are a particular nuisance to frogs. Although some felines prefer to ignore them (even after their jumps elicit the 'chase me' response), many are undoubtedly a significant threat to local amphibians, sometimes gnawing upon their legs while the frogs are still alive.

Many predators, including Otters (*Lutra lutra*), make the most of the sudden appearance of frogs and toads when they are on their breeding migrations. To avoid the poison glands of toads, Otters 'peel' the skin off the torso and legs, often leaving these skins in a neat little pile by ponds and rivers while they feast on the meat underneath. Human passers-by may be confused and disgusted by this – as with the crow predation in Germany – but in many respects it is all very much the amphibian way: boom and bust. Some years will be good for toads and some years (when predators abound) will be bad. By producing so many eggs at a time, surviving frogs and toads are capable of restocking populations quickly after particularly bad years, so this phenomena is normally nothing to worry about too much.

Below:

Snake in the grass

One of the oldest foes of frogs and toads across most of England and Wales is the Grass Snake (*Natrix helvetica*), which is apparently immune to toad poison. So deep is the evolutionary history between this predator and its prey that, upon seeing a snake, both frogs and toads immediately become agitated. Frogs will leap away, trying to gather distance between themselves and the slower-moving hunter. Toads, far slower than their frog cousins, opt for a different behaviour. They stand on the tips of their toes and inflate their bodies with air, attempting to look as large and ungainly as possible. For some Grass Snakes, this behaviour may urge them to reconsider their attack. Swallowing large objects takes time after all, and no snake wants to face the indignity of having to regurgitate a prey item that is too unwieldy to swallow. Plus, with their mouth full, Grass Snakes are themselves open to attack from predators. For them, each prey item needs to be considered carefully. The toad, meanwhile, must hold its nerve.

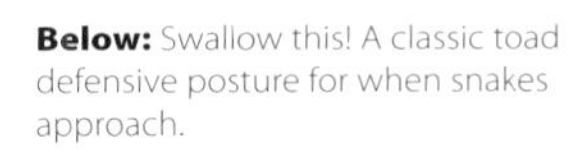

Below: Swallow this! A classic toad defensive posture for when snakes approach.

High-stakes lottery

Many adult frogs and toads are graduates of that most dangerous arena of all: the pond. These waterbodies abound in tadpole-eating predators, including fish, Greater Water Boatman (*Notonecta glauca*), Water Scorpions (*Nepa cinerea*), newts, and the carnivorous larvae of dragonflies, damselflies, mayflies and, particularly, water beetles. The stakes are weighted heavily against frogs and toads in this, their early life stage. In some years, only a handful will ever make it to metamorphosis.

A well-worn estimate for British frogs and toads is that just 1–5 per cent of tadpoles will make it out of the pond as metamorphs, and of these, only 1–5 per cent will make it to breeding age at two or three years. In other words, for every adult frog or toad you see, there were perhaps a thousand tadpoles that didn't make it. But even as adults, the chances of survival each year are skewed towards ill fortune. In the UK, it is estimated that, for every year of life on land, 50 per cent of adult frogs and toads in a population won't make it back to the pond the following year to breed. Based on numbers like these, it is a lucky frog indeed that will live to five or six years, and a luckier frog still that lives to see a decade.

Defensive duties

Frogs and toads differ quite notably in how they behave when confronted by a predator. In those early moments, frogs may attempt a dash escape courtesy of a series of long leaps towards the safety of deep grass, log piles, rock piles or, should one be handy, a nearby pond or stream. For them, the places where predators cannot squeeze in are worth knowing, and it may be that frogs keep a working 'map' of such local sites in their heads. Toads, however, do the complete opposite when attacked by a predator. Instead, they rely on poison.

Toads produce sticky white secretions in the parotoid glands behind their eyes, which contain a trio of fast-acting molecules. Two of these – bufogenin and bufotoxin – affect the cardiovascular system; the third – bufotenin – is a powerful hallucinogen similar to that found in some mushrooms. Although the poisons produced by many creatures are lethal, toad poison is more of a warning than a death sentence. Animals such as dogs, cats and Foxes that have a go at a Common Toad soon become agitated; they pant and salivate, and become very eager to wipe the disgusting taste from their mouths. If they actually eat the amphibian, however, things can get very serious. Dogs that consume toads, for instance, may start to vomit, and can have palpitations and exhibit extreme lethargy, anxiety and confusion for a few days. Veterinary assistance can be vital and is recommended in all cases. Thankfully, our resident frogs and toads cannot release lethal toxins like those of exotic amphibians such as the Central and South American poison dart frogs.

Above: [illegible]

Producing poison is clearly costly for toads, so releasing it is not something they are particularly eager to do. Many will activate their poison glands as a final roll of the dice, when chewed up in the mouth of a would-be predator, for instance or, in the case of Natterjack Toads, if pecked at by a hungry gull.

Poison glands and leaping legs are not the only tricks that anurans can use to defend themselves. For example, certain frogs, it seems, can scream. When cornered by a predator, some Common Frogs are capable of emitting a high-pitched scream to ward off danger. To human ears, this piercing noise sounds similar to the cry of a baby. The general feeling among scientists is that this is an evolved response that, although delivered infrequently, may serve to attract the attention of other large animals, some of which the predator may itself fear.

Above: The Yellow-bellied Toad (*Bombina variegata*) performing the unken reflex – a weird, but apparently effective, defensive posture.

Occasionally, Common Frogs elicit another strange defensive behaviour, the so-called unken reflex. In this, the frog essentially lies on its back and plays dead when confronted by a predator. Some frogs go further still, acting out a variation of this behaviour that sees them using their forearms to cover their eyes. Although it looks at first as if the frog is comically shielding its eyes from the terror it faces, the behaviour probably serves to remove key stimuli from the predator, encouraging it to lose interest and move on.

The degree to which both screaming and playing dead are inherited frog behaviours remains unknown. Once again, this is an area of study that few herpetologists have yet to investigate, and where weird and exciting new discoveries could one day be made.

Below: Cats and dogs can sometimes fall foul of the toad's poisonous defences. See a vet if you have concerns.

A frog for all seasons

Frogs and toads are often referred to as cold-blooded, but how much truth is there in this catch-all term? The adjective, sometimes used by humans as a slur, suggests sluggish creatures that are totally at the behest of the elements and require warm days in order to thrive. But this is not, strictly speaking, true. All of our British amphibians, for instance, can be active throughout mild winters, many of them bracing weather at which many mammals would baulk. And they use a variety of tricks to modify their temperatures, too, just like we do. Sure, they can be cold-blooded, but on warm days they become, well, warm-blooded.

Zoologically speaking, amphibians are poikilothermic. This means that their internal body temperature varies considerably hour to hour and day to day, and that they have adaptations to make the best of the situations their environment may throw their way. Homeotherms, like humans, are the opposite – our internal temperature is regulated within strict operating boundaries. For us, constancy is key.

Below: [illegible]

One way frogs gather extra heat is by basking. On spring mornings, frogs (particularly Pool Frogs) like to warm themselves in the sun, in much the same way as snakes and lizards, those other well-known poikilotherms. This may serve to speed up their metabolism or (as with snakes) ready their eggs and sperm. For this reason, if you are looking for frogs at the water's edge, it can pay off to scan the sunnier side.

Frogs face a far bigger challenge than toads when it comes to retaining moisture in the summer months. As they have highly porous skin, the water in their body evaporates far more quickly than from a toad of a similar size. This can catch some frogs out. On sunny days, it is not uncommon to see the desiccated carcasses of Common Frogs in the middle of a lawn or at forest edges – these victims strayed too far from the water's edge in search of food and got caught short. Being slightly more thick-skinned, toads don't appear to suffer such indignity. Although the skin on their ventral surface remains permeable, their tough warty dorsal surface probably acts as a waterproof shield.

Above: Common Toads sometimes drink through their skin from newly formed puddles.

Although Britain's frogs and toads like it warm, none like it too hot. On sunny days, for instance, toads often bury themselves in deep, damp soil or mud to limit their exposure to high temperatures. Using their back legs as shovels, they discreetly shuffle themselves backwards into the substrate, sometimes becoming completely submerged. In this respect, toads can enter a state that resembles aestivation in some reptiles, powering down bodily functions to conserve fat reserves until conditions get a bit milder.

Winter challenges

In the colder months, Common Frogs also power down, entering a state that differs from mammalian hibernation only by a matter of degree. Strictly speaking, dormancy is probably a better description. After all, unlike the vast majority of hibernating mammals, amphibians can be roused from their winter sleep, particularly on warmer days, when they might emerge for a few hours or so for a spot of foraging. Compared to toads, frogs are particularly prone to this sort of wakeful behaviour. Male Common Frogs – many of which sleep through the winter on the pond bottom – may move solemnly around the pond or bob for a few moments at the surface before swimming back to the bottom. During icy periods, they may even be seen through the frozen pond surface, moving around like ghostly figures underneath. Interestingly, Pool Frogs seem to prefer terrestrial places in which to see out the colder months.

Like frogs, toads also enter a deep state of inactivity in winter. By November, most will dig down into soils (or sand in the case of Natterjacks) and do not appear again until early spring. Keen gardeners sometimes come across Common Toads in this state when digging. Gently putting them back is probably the best advice, ensuring they are deep enough to escape frosts and the interests of hungry neighbourhood Foxes and cats. Winter is clearly an important time for toads. Research has shown that in milder winters they burn up more calories than in colder, longer winters, emerging in spring with fewer fat reserves. If sustained mild winters become a common phenomenon in the UK and Europe, it is unclear what impact this may have on breeding populations of toads. In some cases, this lack of adequate dormancy may be yet another stress contributing to local population declines.

Above: [illegible]

One surprising fact about Common Frog tadpoles is that, if conditions aren't quite right in the pond for metamorphosis to take place, they can slow their development and overwinter there. How often this happens in the wild is unclear, but it certainly occurs in some garden ponds – my own included! In fact, a 2004 survey in Glasgow suggested the behaviour was relatively common, with overwintering tadpoles occurring in one in five garden ponds. The mechanism through which tadpoles come to delay development is of interest to herpetologists. Research indicates that some tadpoles seemingly 'decide' as early as July to slow growth and focus instead on overwintering. It may be that these individuals benefit by metamorphosing at a larger size the following year or emerging earlier than their rivals, giving them a chance to make more of the spring glut of food.

Winterkill

Being highly porous, frog skin offers more than just protection from the elements. Interestingly, it can act as a boundary through which gases can pass – in effect, acting like a lung. Male Common Frogs make full use of this feature, choosing to lie submerged in ponds during the winter months, perhaps as an adaptive behaviour to guarantee their place at the grand reproductive spectacle that comes in spring. There, at the bottom of the pond, the male frogs bide their time, allowing oxygen to diffuse into their bodies while carbon dioxide escapes. Although impressive, the behaviour does come with risks. In cold winters, for instance, when ice covers ponds for many days or even weeks, pond owners become familiar with the depressing sight of bloated frog carcasses trapped under the ice – the victims of so-called winterkill. These are individuals that have, sadly, suffocated.

One way to limit winterkill in frogs is to try to ensure that dissolved oxygen levels do not dip during periods when the pond is frozen. The Freshwater Habitats Trust (which is currently researching the phenomenon) advises that pond owners gently sweep snow off the frozen pond surface to ensure that light can continue to enter the pond, thereby allowing aquatic plants to continue photosynthesising. Dissolved oxygen mixes slowly in water, so it may be that ponds that are deeper than they are wide (and perhaps with lots of leaf litter) are particularly susceptible.

Below: A sad victim of winterkill – a winter phenomenon that is finally getting the research it requires.

Reproduction

Frogs and toads are termed explosive breeders, because their sex lives play out in a matter of days each year. In spring, they begin by making migrations to breeding ponds, in which they battle and writhe to gain the interest of a mate. All the spare energy frogs and toads have saved throughout the year is for this moment, so the pressure is on. As in all aspects of life, not everyone will be a winner.

The best time to observe frogs and toads is when their minds are on other things, especially reproduction. While they are preoccupied with their seasonal struggles, you can watch them for hours and barely be noticed, so focused are they on their goal of finding a suitable mate. What we see in their reproductive behaviours at this time of year is a hidden world of politics, rage, lust and corruption; a place of intense energetic interchange and high-stakes gambling; a world that rings with a patterned chorus to those who listen and a cacophony to those who don't. Here, we consider in detail the sex lives of our native frogs and toads.

Opposite: Common Frogs in amplexus. Female beneath; male on top.

Below: Amplexus – the go-to sexual position for many (but not all) anurans.

On the move

Right: [illegible]

First, let's look at that most splendid of amphibian intuitions: migration. How is it, exactly, that frogs and toads find their way back to ponds each spring? For many centuries, their impressive knack for navigation was deemed almost magical. Did they read the stars to find their breeding ponds? Did they follow ley lines? Or was it some form of unseen witchcraft?

The answer was none of these things. Instead, it appears that our native anurans follow their noses. It turns out that frogs and toads are particularly responsive to the scent of the blooming algae their tadpoles will rely upon for nutrition. (And how did scientists learn this fact? Well, let's just say that the use of tiny nostril plugs was involved.) Yet it seems that smell alone can take frogs and toads only so far. Once the adults near the pond, other senses kick in: they use their ears to home in on the croaks and chirps of their respective species, and they use their eyes to calibrate their movements, avoiding obstacles along the way. In the case of Natterjacks, which often depend on temporary ponds that may be present for only a matter of months, the sound of the male's call is a crucial determinant of where the best ponds are, and females are drawn to their whirring calls.

Interestingly, the moon may play a part in migrations too. At some European study sites, a full moon coincided with a greater number of arriving amphibians. In addition,

Below: [illegible]

Above: Natterjack Toads can be heard from many hundreds of metres away.

frogs and toads may also rely on some form of simple memory to find their way. During the years I spent running a frog helpline, I was often struck by how many homes still received visits from frogs each spring years after ponds had been removed. Research undertaken in 2014 confirmed that actively foraging tropical amphibians are capable of using local geographical landmarks to navigate, and this could also be the case for our amphibians. Far from being mindless biological automatons, frogs and toads are clearly highly sensitive to a range of cues that lead them to potential breeding sites. And with 150 million years of evolution on their side, what else should we expect?

Distance and timing

In terms of distance migrated, there is a big difference between our two most widespread anurans. Some Common Frogs, it seems, travel up to 400m (440yd) or so in early spring, and it is at this time of year that you may see them leaping through habitats not traditionally associated with the species, including car parks and city streets. Blessed with long legs and a tendency to leap rather than walk, migrating Common Frogs face little threat from road traffic. Common Toads, on the other hand, are another story.

Being endowed with a more terrestrial body means that Common Toads can roam much further from water than frogs. At some sites, Common Toads move more than 1km (0.6 miles) from overwintering sites to breeding ponds. This, along with their slower mode of travel, means they are far more likely to come into contact with traffic. Indeed, roads are likely to be one reason why the Common Toad is the fastest declining of

Above: Toad crossing sites are helping scientists gather important information about national declines.

Above: [illegible]

Britain's widespread amphibians. Thankfully, networks of 'toad patrols' – local volunteers eager to monitor and move toads safely across roads – exist at many sites (see page 111).

The time at which the amphibians in your local area migrate and spawn largely depends on where you live. If you live in the far south-west of Britain, you can expect frogspawn as early as January. Conversely, if you happen to live in eastern Scotland (or East Anglia), you may have to wait until April in some years, particularly if there is a prolonged dry period or period of late frost. On the whole, Common Toads appear in breeding ponds a week or two later than Common Frogs, although there is often some degree of overlap. Natterjack Toads take things a bit slower, often beginning their spawning migrations later in April, and then come the Pool Frogs, which begin their breeding migrations in May. For both Natterjacks and Pool Frogs, there may be a 'second sitting' of breeding that occurs well into early summer.

Picking the right pond

A pond is more than just a water-filled hole. Each is its own city, with its own foundations, and its own networks of plant and animal inhabitants, and each has its own telltale chemical make-up and physical signature. Our frogs and toads, it seems, know this. Common Toads seem to prefer big ponds or small lakes, often coexisting with fish that seem, wisely, to avoid their poisonous tadpoles.

Common Frogs, on the other hand, are far less picky and many will spawn quite happily in garden ponds, temporary meadow ponds or even large puddles. It is likely that there is an evolutionary advantage to this behaviour, since temporary ponds rarely have fish or large invertebrate predators that may consume their tadpoles. However, the strategy is far from foolproof in smaller ponds and, in some dry years, the sight of a rapidly evaporating puddle filled with the writhing bodies of hundreds of tadpoles can be distressingly common. Still, in nature, reproduction is all about risk. One good year is all it takes for this behaviour to persist, and for some Common Frogs, puddle ponds are a punt worth taking.

Left: Being small or medium-sized and often lacking in fish, garden ponds can be very attractive habitats for spawning Common Frogs.

A frog in the throat

Above: Natterjack Toad calls are a mélange of hidden signals indicating species, sex and size.

Once frogs and toads have migrated to their breeding ponds, they must locate and find members of the opposite sex somewhere in the surroundings. To limit problematic inter-species encounters, each of Britain's anurans has its own call signs, unique to the species. As with the songs of birds, it may be that these calls perform more than one purpose. First, each individual call occupies a frequency that cuts through the amphibian melee, highlighting clearly the right species to the right female, rather like a flag. Second, females can identify the size of males by the depth of their call, which is deeper the larger the male. Male Common Toads also use calls in a third way, as a signal to other males to stand down should, for instance, sexual advances be made.

Interestingly, there is a suggestion that some amphibian species listen in to the calls of others. There is good evidence, for example, that Smooth Newts (*Lissotriton vulgaris*) can home in on the call of the Common Toad as a way to locate a suitable habitat within which they may coexist. In nature, such co-opting of one another's calls is rare, so it is quite a surprise to see it in a species so close to home.

Even with each species piping out its distinctive calling cards, mix-ups are inevitable. On many occasions I have come across frogs attempting to grab toads and vice versa. More concerning still is when frogs and toads grab onto prized fish, to the dismay (and occasional frustration) of koi carp owners in particular!

Below: Occasionally other species, including fish, become the objects of momentary unwanted attention from male frogs and toads.

Sexual strategies and behaviours

The sexual struggles of male and female anurans are demanding but not always identical, with each sex often having its own unique set of challenges and competing rewards. For instance, one could argue that males have a particularly difficult time, since there are more of them. Because males mature more quickly and tend to hang around after breeding, there are nearly always more of them at breeding ponds than there are females. There is more competition between males, and there is more at stake. Yet one could equally argue that the females have it tough: each female may be harassed by an inordinate number of lusty males, after all. In fact, in many species, females can succumb to the attention, drowned in the mating frenzy that ensues. The truth is that the stakes are high for amphibians of both sexes at this time of year. Many simply end up becoming 'spent', rather like the Pacific salmon that migrate from the sea up rivers, where they then spawn and die. Anuran corpses can litter some well-occupied ponds once breeding has finished, a fitting epitaph to the evolutionary struggles that the season brings with it.

Below: Ponds are often home to many more male frogs than breeding female frogs meaning competition between the males can really ramp up.

Left: Many female frogs and toads have plenty to gain by letting males compete for pride of place on their back.

Below: The nuptial pads of some anurans are also able to inject chemical stimulants involved in sex. This includes our native resident, the Common Frog.

Loving embrace

With all this death, it is no surprise that male and female frogs and toads have evolved a suite of adaptations to help them get the best of the breeding season in terms of transmitting their genes to future generations. Among the easiest of these adaptations to observe is amplexus (Latin for 'embrace'), a kind of power hug given by the male to the female to improve the chances that he, and not another male, is the one to fertilise her eggs. During amplexus, the male locks his arms around the female, normally underneath her forelegs. In this position, he lines up his cloaca to release sperm in such a way as to maximise fertilisation of the eggs that the female extrudes when she finally deems the time is right. Males are aided in this embrace by sticky suckers on their thumbs, called nuptial pads, which help them to hold on to females especially tightly.

Once frogs or toads achieve amplexus, the males may be challenged by rival males that try to pull the pair apart with strategic kicks and tugs. In this way, females may gain from mating with the strongest individuals,

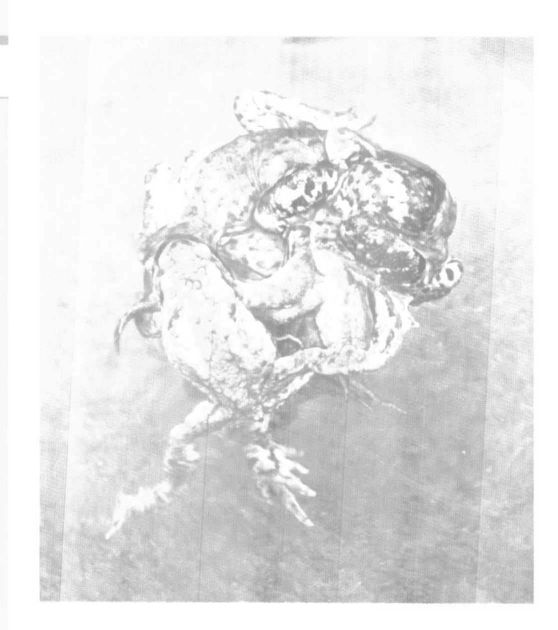

which presumably are also the fittest, reproductively speaking. In effect, the gene pools of local populations become dominated by those healthy individuals that have a knack for endurance.

Sneaky mating

Male and female anurans occasionally manifest behaviours that have more than a hint of trickery about them. In Natterjack Toads, for instance, small males can adopt a 'sneaky' mating strategy. By sitting silently next to a large, dominant male, a small male can make use of its rival's attractive booming call, which is more attractive to females than its own higher-pitched call. On the approach of a female, the smaller male attempts to seize the moment, as if hoping that the female will not notice his diminutive stature. This strategy often fails, but sometimes... well, it only has to work once for an inherited behaviour to persist in a gene pool. In this way, anurans have hit convergently on an impressive evolutionary strategy that is employed by creatures of many taxonomic groups, including beetles, wasps and cuttlefish.

Above: [illegible]

Right: [illegible]

Variety is the spice of life

The weather patterns that occur in the run-up to the breeding season contribute enormously to the speed and intensity of activity at breeding ponds. In some years, amphibian migrations can stall if the nights remain cold and dry for a long time, for instance. When warm, muggy nights do finally arrive, frogs and toads almost seem to burst out of the landscape in a frenzy. Other years are very different, however. I have known seasons where long frosty nights have continued throughout a breeding season, never giving way to rain. As a result, amphibian breeding in these years becomes spread out over weeks, as females arrive in dribs and drabs. Needless to say, the mating frenzy in these seasons is rather subdued.

Above: A lone male Common Frog scans the pond edge for any late-arriving females.

The unique character of each breeding season probably also impacts on the reproductive success of different males and females in the population. In breeding seasons where females converge on a site all at once, for instance, smaller males might be lucky enough to hang on to a female without being usurped by a bigger rival, simply because the attentions of bigger males are being occupied elsewhere. Conversely, in years where the action is drawn out, dominant males have the time and space to defend each and every visiting female. In these years, the smaller males are likely damned. This varied pattern is yet another aspect of amphibian biology that makes these creatures so interesting.

The Life Cycle

Frogs and toads are united in having a familiar tadpole life stage. Although we take this for granted, if you look closely at developing tadpoles you will see a host of anatomical changes that maximise their survival potential. The tadpole stage is arguably the most dangerous in the life cycle of frogs and toads, and only a lucky handful will make it. The rest become food for an aquatic army of predators.

The spawn identity

Sometimes, several days pass with male and female frogs and toads in amplexus, both awaiting some unknown environmental cue as to when egg-laying should begin. Why this delay occurs is unclear. One would think that, once attached, the female would get straight to it, but no. It may be that female anurans are waiting that bit longer before depositing their spawn, allowing the algae that will feed their offspring further time to blossom. Or it could be that they are waiting for other, more suitable, males to appear on the scene. Either way, come spring most amphibian enthusiasts will eventually be marvelling at the familiar jelly-covered dots in their ponds. This is an exciting time, as it is when the numbers game truly takes off. The dice has been rolled, yet the outcome still remains uncertain.

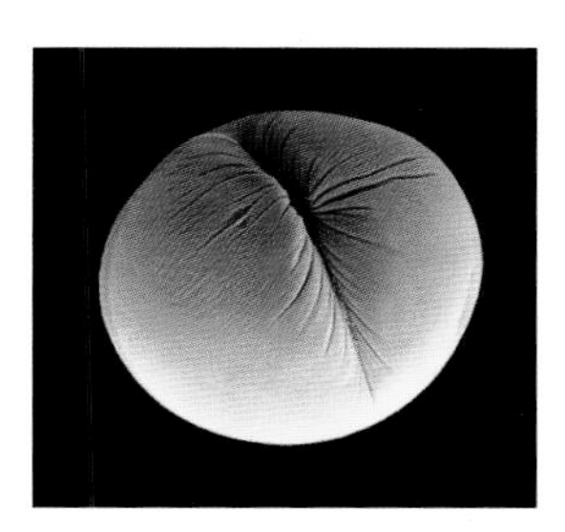

Above: It all starts with a single cell: frogspawn and its subsequent development has helped inform the science of embryology for many centuries.

Frogspawn

When Common Frogs first lay their egg mass, it looks surprisingly pathetic – in many cases, little more than the size of a jacket potato. It hangs in the water for a few days, glued to rocks and pond vegetation, or (often) held in place by the mass of other spawn blobs. During this period, courtesy of a protein that covers each egg, frogspawn absorbs water, which makes the tiny eggs swell to a more recognisable size. If you look more closely at the individual embryos (the little black dots) at this time of year, you can see that they are actually

Opposite: Spawn is often laid for days or weeks. The spawn in the top-right of this photo is freshly laid and has not yet swelled to full size.

two-tone, with an obvious darker patch on the topmost side. There are likely to be a few reasons for this. First, the difference in colour is caused by a pigment, melanin, which serves to protect the developing eggs from damaging ultraviolet radiation. But this coloration may also encourage the eggs to absorb more heat from the sun's light, permitting faster growth. In addition, the two-tone shading of eggs may make them harder to spot by predators below – similar examples of this so-called counter-shading pattern are seen in larger sharks, dolphins and whales.

Toadspawn

Toads tend to spawn in slightly deeper water (30cm/12in or so) than frogs, and their eggs are easy to tell apart. They lay jelly-like spawn in long strings rather than the big blobs of frogs, and they wrap these around pond vegetation. The egg strings are laid relatively slowly by the female, emerging bit by bit. The male toad, gripping firmly on the female's back, feels with his toes to assess when the female is laying and when she is resting, and times the release of his sperm accordingly. One interesting difference between Natterjack Toads and Common Toads is noticeable when observing their ribbons of spawn. If you look closely at Common Toad spawn, you will see that it is actually made up of two chains of eggs; in Natterjacks, it is a single string.

Below: [illegible]

Left: Natterjack toadspawn. Note the single chain of eggs.

Left: Common toadspawn with its distinctive double chain of eggs.

Too much of a good thing

Some people express surprise at how much frogspawn turns up in their garden ponds in early spring. Even ponds that many would consider little more than a water feature can be overcome by the stuff, with eight, nine or even more blobs deposited in some years. Is there such a thing as too much frogspawn? The answer depends on your perspective, really. For frogs and toads, there probably is such a thing. After all, if there are many tadpoles, there are more mouths to feed and therefore more competition between each of them for resources. This is bad news for tadpoles. Intense competition for limited resources can influence and encourage a behaviour we humans consider ghoulish, to say the least: cannibalism. Tadpoles can be adept at bullying, harassing and otherwise making meals of weaker conspecifics and, kept without care in tanks in the corner of a classroom, for instance, often display such a tendency.

For this reason, having lots of spawn in your pond doesn't always mean that your garden will become overrun with plagues of tiny frogs later in the year. It just means that the tiny frogs that do emerge will be the toughest of the batch, individuals whose development was partly powered by, well, siblicide – nature, red in tooth and claw, as the poet Tennyson put it. Interestingly, there is some evidence that tadpoles can show adaptive responses to overcrowding, leaving the pond as smaller metamorphs when competition for food becomes more intense.

Below:

One thing is clear: if you have lots of spawn, it's unlikely to last. Anuran populations are renowned

Spawn swapping

Spawn swapping was one of those idiosyncratic behaviours nature lovers once partook in, whereby those with too much spawn would donate it to schools and neighbours with ponds bereft of amphibian activity. My first memory of amphibians was just such an encounter, when a washing-up bowl of frogspawn was gifted to our school by a local villager. Sadly, such spawn swapping is no longer advisable, for a couple of reasons.

First, moving buckets of pond water around the country is a risky practice in itself, since it can accidentally transfer invasive species of aquatic plants from gardens into wild habitats. There, they often flourish, outcompeting native wildlife, proliferating and, in extreme cases, causing local flooding.

Second, there is a very real possibility that amphibian diseases (see page 77) can be spread by translocating spawn. At least one common disease of amphibians, an American ranavirus, has made it into the wild in the UK, probably through the pet trade. It is a disease that can be passed into new populations through the movement of infected spawn. For this reason in particular, moving spawn around is not a good idea. If you have a pond without spawn and you're desperate for anuran visitors, hold your nerve if you can – in many cases, they will come. Common Frogs are quick to colonise new places for breeding, and even some very urban ponds can become haunts for the species if they are left alone for a year or two.

Above: Newts, including the (non-native) Alpine Newt (*Ichthyosaura alpestris*) are one of many known spawn-eating predators.

for fluctuating quite wildly at times, cycling over years and decades. Some years will be good; other years will be bad. My advice if you have an excess of spawn in your pond is to make the best of it, without worrying too much.

Egg development

And so, back to those individual eggs. Within days, what was just a couple of cells becomes a ball of cells millions strong. The black dot in the jelly elongates. A tail bud forms, then something you could almost call a head, then gills and a heartbeat. If you watch the embryos closely at this point, you will notice that the comma-like immature tadpoles wriggle faintly within the jelly. They are on their way.

These early days are probably the safest moment in an amphibian's life, as only a handful of creatures can pass through the jelly force field to feast upon them. Spawn-munching predators include some flatworms, Moorhens (*Gallinula chloropus*) and newts, the latter thrusting their bodies at the mass of spawn open-mouthed, like sharks stripping a dead whale. After a few more days, or up to a week or two depending on the weather, the tiny tadpoles begin to migrate through the spawn, resting in a little pool at the top of the mass of jelly in the case of frogs. Here, the siblings spend time feeding and making the most of the warmth in apparent safety. More days pass, and slowly the jelly begins to sink. A new stage begins: the tadpoles are now free swimmers, and are at their most vulnerable.

Tadpoles – a numbers game

Life as a tadpole is a lottery, so numerous are the predators that get fat off this plentiful food source. Not only do tadpoles have to escape from fish, newts and birds, they also have to dodge a volley of threats from underwater invertebrates, in particular beetles and their larvae. Greater Water Boatman, Water Stick Insects (*Ranatra linearis*), Saucer Bugs (*Ilyocoris cimicoides*) and Water Scorpions also make short work of tadpoles. These nightmarish predators use their straw-like mouthparts to suck tadpoles dry, making the most of the nutrients within. Dragonfly nymphs, with their lunging, grabber-like mouthparts, also have a certain fondness for tadpoles, as do Water Spiders (*Argyroneta aquatica*) and some leeches. Perhaps the most notorious of tadpole predators, however, is the Great Diving Beetle (*Dytiscus marginalis*), whose adult and larval stages are extraordinarily adept at seeking them out. A single adult Great Diving Beetle may go through as many as 20 tadpoles in a day, and I have even seen their larvae attack newts and sticklebacks!

Not all ponds have such an illustrious list of predators, of course. Frogs and toads that locate new ponds, or happen to spawn in ponds that have dried out and then refilled, and so are

Above: With large eyes and mouthparts able to slide out from [illegible] dragonfly nymphs are [illegible] adapted to feed upon amphibian larvae.

Right: Some large spiders, including the Fen Raft Spider [illegible] capable of [illegible] tadpoles, which they drag to the surface to eat.

almost devoid of predators, momentarily hit the jackpot in terms of the young metamorphs they give rise to. These are perfect ponds for amphibians, but there are many that are far from ideal. I have seen whole ponds emptied of their tadpoles by predators that are simply too good at finding and feasting upon them. Life is hard for tadpoles, which is why anurans have evolved to produce so many. It's a numbers game, after all.

Tadpole diets

Turning back to the survivors, the young free-living tadpoles start off as consumers of algae and single-celled protozoa, and they often gather in the parts of ponds where these organisms aggregate. Avid frog watchers may notice, for instance, that on sunny days tadpoles are easy to observe grazing upon sunlit rocks at the water's edge. In a matter of weeks, however, their diet alters as their intestines shorten. At this time, their tastes move towards larger food items, including dead animals and, as discussed, their tadpole pond-mates, which they soon home in on should they succumb to weakness or starvation.

The simple mouthparts of tadpoles, contained within an oral disc, are surprisingly monstrous when seen in close-up, comprising sharp keratinous beaks within which sit rows of tiny keratinous teeth. So unique to each tadpole species are these oral discs that their arrangements have become a diagnostic tool for some scientists trying to differentiate species of amphibian larvae in the wild.

Above: If forced into low nutrient environments, tadpoles regularly bully, kill and consume their siblings.

Left: Up close the horny beak of a Common Frog tadpole gives it a certain shark-like charm.

Tadpole identification

Above: The jet-black tadpole of the Common Toad.

Above: Natterjack Toad tadpoles are often confused with tadpoles of the Common Toad.

Thankfully, in the UK at least, the larvae of our widespread frogs and toads aren't too difficult to tell apart. As they grow, Common Frog tadpoles are mostly brown with golden speckles that become apparent when the sun is bright. Pool Frog tadpoles are slightly more olive-green, with a vaguely pinkish underside. Both Common Toad and Natterjack Toad tadpoles, on the other hand, remain jet black throughout their development. Size is another diagnostic aid for identifying tadpoles. While the bodies of Common Frog, Common Toad and Natterjack Toad tadpoles grow to about the size of a baked bean (excluding the tail), those of Pool Frog and Common Midwife Toad tadpoles become noticeably larger and more like, say, a broad bean. Things get a bit harder when trying to tell Common Toad tadpoles from those of the Natterjack Toad. In these cases, even the experts can be caught out.

Interestingly, tadpoles of varying species also differ in their behaviour – so much so, in fact, that this can be another diagnostic tool in your amphibian-spotting arsenal. If you see shoaling behaviour in tadpoles, for instance, you are probably looking at the larvae of a Common Toad. Likewise, if you see tadpoles swimming apparently nonchalantly in the middle of the pond, then these are likely to be toad tadpoles. The reason for this is that toad tadpoles possess the same poisons that make the adults so unpalatable to predators, and are therefore less troubled by the presence of newts and fish.

Above: Tadpoles of the Pool Frog measure up to 70mm (2.8in) long.

Above: The Common Frog tadpole. Note the spots along the body.

Changing times

Above: The Common Frog provides one of the world's most beloved life-cycles. In all, it takes 12–16 weeks to go from egg to tiny metamorph.

In the final developmental stages of frog and toad tadpoles, a suite of visible changes – termed metamorphosis – occur as they ready themselves for life on land. In addition to the obvious appearance of a pair of back legs, the eyes now begin their journey to the top of the head and the mouth becomes widened along the front arch of the face. The tadpoles begin to look less spherical now, probably because their intestines – once long in order to cope with a herbivorous diet – become shorter in readiness for the wholly carnivorous diet that will follow soon after metamorphosis.

There are other changes too. Two strange lumps appear on either side of the body behind the eyes, indicating that the front legs are growing inside the body. Unlike the back legs, these forelegs develop internally before punching through a specially weakened 'window' within the walls on either side of the tadpole's body. The first punch is often through the spiracle, a pre-existing respiratory hole, then comes the other, at which point the tadpole has four legs. In this state it immediately seems to behave more like an adult, even though its long tail remains. It clambers around at the water's edge and regularly appears at the surface to eye up the terrestrial world above.

Above: Its eyes firmly on top of the head and its jaws wide, this Common Frog is likely to complete its metamorphosis in a matter of days.

But that is not the whole story. In the last section of the body, an impressively coordinated cellular death (called apoptosis) continues. The same process through which the tadpole's body walls were weakened to allow the forelimbs to pop out also causes the dead cells in the tadpole's tail to be reabsorbed and recycled. As if by magic, the process of metamorphosis nears completion. In a matter of weeks, the apparently lifeless clump of cells in a jelly egg has become a thumbnail-sized hunter, a beautiful, charismatic survivor. Its terrestrial lungs are ready, and the metamorph walks out of the pond: one small step; many giant leaps.

All in good time

So, how long does it take to make a frog or toad? The answer to this popular question depends partly on the species and partly on the pond in which the tadpoles develop. In general, for the main British species at least, the period from egg-laying to metamorphosis may take between 10 and 16 weeks, with metamorphs leaving ponds during damp spells in May, through June and sometimes into July. In Natterjacks, which are adapted to more temporary ponds, development is far quicker – occasionally as little as six to eight weeks. In all species, it can help to look at the pond to pinpoint exactly when metamorphosis might happen. Of the local factors limiting development in the water, food availability and temperature (themselves linked) are especially important. On the whole, shady ponds give rise to metamorphs later than exposed ponds. Likewise, upland ponds are more likely to generate metamorphs later in the year than lowland ponds.

First steps

So, now we have our froglet or toadlet, a terrestrial world-beater. To survive to breeding age, this tiny creature must first chisel out a life for between two or three years on land. Once again, the odds of survival are, predictably, stacked against it. In these early days, the most pressing threat to metamorphs comes not from predators or disease, but from the sun. The drying effects of the sun on terrestrial organisms bring immense problems for froglets in particular. Their tiny bodies, wrapped in semi-permeable skin, have an extraordinarily large surface area to volume ratio compared with adult frogs, which means that they can expire very quickly, dehydrating within minutes of escaping moisture-rich environments on a hot day. For this reason, froglets and toadlets often stay put at the pond edge for many days and weeks during sunny weather, waiting for a rainy day (or night) during which they can make their dash for cover safe from the elements.

Above: Common Toadlet metamorphs gather on top of a patch of floating grass. Many will wait until the summer rains come to make their dash from the pond.

Once the summer rains come, the little amphibians radiate out of the pond in all directions, in search of suitable damp habitats. For naturalists, this is quite a wildlife spectacle. In fact, in some years there may be so many tiny anurans on the move that the ground seems almost alive with them. It is not long, of course, before the predators catch on. Magpies (*Pica pica*), crows and Blackbirds (*Turdus merula*), can become attracted to the melee, as can curious pets including chickens. At night, Hedgehogs, Foxes, Badgers and rats join the ranks. The sight of hundreds of froglets or toadlets being swallowed by these predators is troubling to some pond owners, but, on the whole, this is largely natural. However, if you have a cat or a dog, this might be a good time of year to keep it indoors.

The froglets and toadlets that escape the clutches of predators often gather in damp spots underneath garden objects where woodlice and other smaller invertebrates abound, and a quick check under plant pots, logs and rocks may reveal some. Although it may be tempting to pick up a metamorph, this is not a good idea. Our skin, dry to the touch, may encourage the diffusion of water out of their little bodies, stressing them into an early death. Such an action is also likely to be ten shades of terrifying to a small creature like this.

Even after years of study, scientists still know very little about the behaviour of froglets and toadlets during this crucial time in their life cycle. They certainly appear to put on weight quickly, doubling their size within weeks

Below: In the early days soon after metamorphosis, young froglets and toadlets often gather together in nearby cracks and holes while waiting for summer rain.

and then again within a month or two. Ants, tiny worms and small spiders and beetles probably make up a large part of their diet at this point, but they have a long way to go. Frogs and toads may remain in this infant state for two years, and female toads sometimes for three years. This extra year gives females more time to get to a healthy breeding size.

Above: Radio-tracking studies of the Polecat (*Mustela putorius*) in Europe show that amphibians are an important part of their diet.

Left: As their tails are re-absorbed, frog larvae begin to act more like adults, regularly hopping about and prowling the water's edge.

Under Threat

Amphibians are one of the fastest disappearing groups of animals on Earth, with a third of all species currently facing extinction. But how did this situation arise, and what hope is there for UK species? This section investigates the causes of the decline in global and local amphibian populations, from deadly diseases and parasites to environmental and other human-induced threats.

Parasitic enemies

Opposite: Has this Great White Egret (*Ardea alba*) bitten off more than it can chew? A variety of parasites live in frogs, and many are capable of infecting birds.

Frogs and toads are rich in parasites. These organisms feed on (and within) many amphibians and some can hinder their reproductive success or, in some cases, lead to death. While most parasites have little impact on populations overall, some have been implicated in catastrophic declines.

Many of the organs of frogs and toads have become island ecosystems for a host of unusual worm-like parasites, about which scientists are still learning. Flukes and nematode worms, for instance, haunt the anuran digestive system, *Polystoma* worms occupy the urinary passageways, and some parasitic worms even make their home in the eyeballs of frogs and toads. Ectoparasites (external parasites) include monogenean flukes and leeches (including fish leeches), which are commonly seen attached to adult amphibians (and even tadpoles), as well as mites and ticks during the terrestrial phase. Some adult frogs and toads regularly suffer from chiggers, the larvae of mites that burrow deep into skin, causing inflammation.

In North America, worm-like trematodes in the genus *Ribeiroia* inflict a very unusual phenomenon on frogs: they cause them to grow extra legs. The life story of these strange flatworm-like parasites is fascinating. As larvae, the trematodes spend their early days inside the bodies of freshwater snails, where they clone themselves into an army that later bursts forth from the snail into the pond.

Below: For years, six-legged frogs were considered harmless mutations. The truth has turned out to be far grizzlier.

At this stage they become free-swimming larvae. Their quarry? Tadpoles. The tiny larvae latch onto tadpoles – specifically the hind parts, where the legs develop – at which point they secrete a chemical that causes limb deformities, namely the growth of extra legs. The tadpoles, should they metamorphose successfully, then become adult frogs with extra legs. But why, you might ask, would the trematode want to create a frog with extra legs? Well, being unable to hop properly, the parasitised frogs are an easy target for birds, including birds of prey and herons, and this is where the trematode wants to be. Incredibly, the only place this tiny worm-like creature can breed and lay its eggs is inside the body of a bird. In other words, it uses the frog as a vehicle to get into a final bird host in which it can have sex. Natural selection appears adept at building strange parasitic relationships like these, although not all quite so convoluted.

Above: A Common Toad plugged up with parasites of the Toad Fly.

Common Toads are no stranger to unique associations with parasites. In the UK and across Europe, the species has its own personal parasitic blowfly: the Toad Fly (*Lucilia bufonivora*). Each year, these ordinary-looking insects seek out particularly large toads upon which to lay eggs. On hatching, the Toad Fly larvae crawl in through the toad's nasal passageways and begin feasting on the internal tissue and brain. Late in the summer, when temperatures start to dip below 14°C (57°F), the maggots emerge en masse from the toad, killing it in the process, before pupating in the soil. It's all spectacularly gruesome. In some summers, perhaps 10 per cent of adult Common Toads are infected with the parasite, and each and every one will die as a result. In fact, toads probably have more to fear from Toad Flies than they do from any large predator, snakes included.

End of an era?

Smaller organisms – namely fungi, bacteria and viruses – can also parasitise adult frogs and toads and their tadpoles, sometimes with devastating effects. Today, many amphibian populations around the world are collapsing, and scientists are working quickly and desperately to understand the causes before it is too late. Recent statistics on amphibian declines speak for themselves: at the time of writing in 2018, amphibians are facing extinction at a rate 211 times greater than would be expected for their taxonomic class, with 2,100 of the 6,966 known species at risk of dying out. This makes amphibians one of the most threatened groups of animals on the planet. For them, the clock is ticking.

Above: *Telmatobius* – a frog genus native to the Andean highlands of South America – is one of many anurans moving closer towards extinction because of chytridiomycosis.

What could have caused such shocking declines? One common phenomenon across many tropical amphibian populations has been sudden mass die-offs, even in pristine sites apparently unaffected by human activity. This was studied most intensively in the Americas and Australia in the 1990s, and the cause was discovered to be the same on both continents. The culprit? *Batrachochytrium dendrobatidis*, a fungal ectoparasite (often referred to more simply as chytrid) that causes death by arresting the amphibian's skin function, limiting its ability to respire and flush away toxic chemicals, and leaving it prone to secondary skin infections. The disease caused by the fungus is called chytridiomycosis, and is perhaps the most serious killer of amphibians on Earth at this time. Frustratingly, new research has suggested that climate change may make the disease even more transmittable across the tropics.

Yet all may not be lost. To describe amphibians as passive victims of chytridiomycosis is to downplay their impressive knack for survival. Not all species are affected by the disease, and there is already evidence that some species are capable of evolving new immune defences against the onslaught. And now that the culprit is known, scientists are working on limiting the spread of chytridiomycosis further while also researching a possible cure. Chytrid fungus is present in some parts of the UK.

Frogs and toads regularly suffer from other infectious diseases. *Ranavirus*, a non-native genus of amphibian and reptile viruses, can cause sudden die-offs in populations of Common Frogs, particularly in the south-east of England, where the disease has become entrenched. These viruses can be transmitted quickly through urban populations, and in some summers hundreds of ponds are affected. Sadly, at the present time there is no cure for ranaviruses, as is the case with chytridiomycosis. Thankfully, however, neither poses a threat to humans, dogs or cats, or any other domestic wildlife.

Below: A victim of ranavirus. Though redness ('red-leg') can be a symptom of infection it is not present in every case.

There is much we have yet to learn about amphibian diseases, so reporting unusual cases of frog and toad mortality has real value to scientists. If you see any unexplained or mysterious frog or toad deaths, submit a disease incident report to Garden Wildlife Health Initiative (www.gardenwildlifehealth.org), an initiative run by the Zoological Society of London and the amphibian charity Froglife, along with the British Trust for Ornithology and the Royal Society for the Protection of Birds (RSPB).

Wild harvesting

Above: Frogs are big business. Here, thousands of frogs harvested from paddy fields in China's Hunan Province are being prepared for trade.

Humans are an oft-forgotten predator of anurans. Although here in the UK we tend to consider frogs' legs a quirky French delicacy, amphibian meat is a vital part of the human diet for many cultures across the world.

Today, Indonesia is one of the world's largest exporters of frog meat, shipping more than 5,000 tonnes (4,900 tons) each year to countries all over the globe. It is cause for concern that much of this produce is not farmed, instead coming from wild-caught frogs. The impact that such wild harvesting may have on populations is still being determined, but its intensity is unlikely to be sustainable in the longer term.

Many frog species are also harvested from the wild for the pet trade and for use in traditional medicines. Sadly, statistics for the numbers of frogs harvested illegally in this way are not known. TRAFFIC, an international non-governmental organisation that lobbies for better management of the international trade in wildlife species, estimates the black market for anurans to be worth hundreds of millions of dollars each year.

Environmental effects

Above: Clean expanses of fresh water are vital to the long-term success of many conservation projects, including the Pool Frog reintroduction scheme pictured here.

Disease is just one of many smoking guns responsible for recent amphibian declines around the globe. Invasive non-native animal species are another threat, squeezing out resident species or bringing in diseases, taking already stressed native amphibian populations closer to collapse. But an even more pervasive threat is habitat loss. The lack of clean, accessible fresh water alongside the terrestrial habitats frogs and toads depend on in their adult life stage is perhaps the greatest danger facing many amphibian populations. However, this can bring us some hope. If we manage our freshwater and terrestrial habitats better, amphibians will flourish and so will we. Just like them, we depend on the same life-blood: clean water. We are different to amphibians, but only by degree.

UK threats

It is humbling to think that almost every wild frog and toad in Britain is related to the early amphibians that migrated from Europe after the last Ice Age 10,000 years ago (see

Below: A truly natural pond: this water-filled depression – a pingo – formed as a result of a giant chunk of ice that melted here 10,000 years ago.

page 25). That post-glacial world of wild streams, oxbow lakes and seasonal pools has changed dramatically over recent centuries, mostly because of our industrial desire to make water work for humans and livestock, rather than for the needs of wild creatures. In the UK today, true 'wild ponds' are surprisingly rare, and in certain parts they are becoming rarer still. The ponds worst hit by human

Above: Leave no trace? Pollution in ponds, including from single-use plastics, has become an ongoing challenge to pond conservationists.

Left: Here, a newly reintroduced Pool Frog takes in the sun. Reintroduction schemes around the world could offer new hope to lost species.

Above: Death by a thousand cuts: within 30 years the Common Toad has declined by almost two-thirds in the UK.

activities were those that once existed on sandy heathlands and coastal dunes, many of which were lost to tourism developments and forestry plantations in the 20th century. Gone with them are the Natterjack Toad (and, in East Anglia, the Pool Frog) populations they once contained. Thankfully, at the time of writing these amphibians and their breeding ponds are protected by UK law. As long as those laws remain, so shall they.

While the UK's Common Frogs seem to have adjusted well to our insatiable appetite for garden water features,

Right: For some UK-wide species, garden ponds have become a valuable lifeline.

Above: A range of wildlife conservation organisations now exist to help the public engage more with the UK's amphibians.

Common Toads, with their preference for larger ponds, have not. Being slower to colonise new breeding ponds, toad populations appear particularly prone to genetic stagnation, especially if they become surrounded by housing and other human infrastructure that limits their movements around the countryside. Once Common Toads are sectioned off like this, all it takes is one bad winter or a rogue pollution event and whole populations collapse without anyone really noticing. This is a particular problem in the south-east of England, where urban development has been most intense. One by one, Common Toad breeding populations have been snuffed out – in 2016, the wildlife charity Froglife announced that, over the preceding 30 years, Common Toad numbers had declined by two-thirds across the UK. In just three decades – little more than a human generation – literally hundreds of thousands of toads have silently disappeared from our lives. A downward spiral has taken place, but it is one we can easily reverse if there is the will to do so. Thankfully, many conservation organisations are attempting to do just this. For more on what you can do to help, see page 108, and for the names and websites of amphibian and other wildlife organisations, see page 125.

将軍太郎平良門
肉芝仙人眷属
吉
伊勢

Frogs and Toads in Culture

Perhaps more than any other group of animals on Earth, frogs and toads have enthralled, entertained and inspired human cultures for centuries, particularly through stories and songs. But it wasn't always this way, for frogs and toads have been on a cultural journey that has seen them go from saviour to supernatural, and from seance to science. In the modern age, the curious relationship we have with amphibians is ever changing.

With their strange life cycle, their charismatic demeanour and the curious chemicals they so often wield, it is no surprise that frogs and toads have piqued the interests of humans for thousands of years. What I find most interesting about them, however, is the transient relationship they have with so many cultures. To every generation, all over the world, it appears frogs and toads are capable of meaning something different. In this way, our relationship with anurans is constantly changing. In this section we explore some of these changes, working our way from deep in human history, when frogs were considered gods, through to the modern day, the age of internet amphibian memes and questionable frog pop songs. Let's start at the beginning, which as is so often the case, is all to do with sex.

Opposite: Often representing a way of life alien to most land animals, amphibians feature heavily in cultural representations from across the world.

Right: A rather charming frog from the 18th-century Japanese work *The Picture Book of Crawling Creatures.*

Sex symbols

As you will have gathered, one of the most charismatic things about frogs and toads is the rather raucous and showy nature of their sex lives. Within the space of only a few nights, ponds can go from apparently lifeless places to scenes of exuberant sexual display and all-out warfare as amphibians make their return en masse to water. It is no surprise, therefore, that some of the earliest references to frogs and toads in human literature feature their sex lives very prominently.

For the ancient Egyptians, dependent on the annual flooding of the Nile for agriculture, frogs were symbols of life and fertility, with their appearance presaging the life-giving inundation. For this reason, anurans came to be heralded and celebrated in the form of the divine

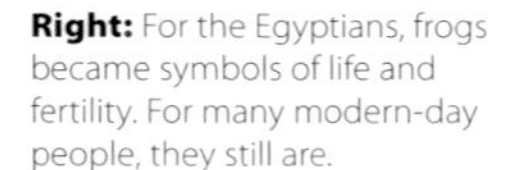

Right: For the Egyptians, frogs became symbols of life and fertility. For many modern-day people, they still are.

Above: After the rains, the floodplains of the River Nile become a warm green – a time for amphibians beckons.

frog goddess Heqet, an amphibian deity privileged to have had her own cult in the Early Dynastic Period, approximately 5,000 years ago. In time, as Egyptian culture moved on and new agricultural techniques and practices developed, the figure and character of Heqet changed. Within centuries, the goddess became reappropriated as a symbol of human fertility and childbirth, and was known as 'She who hastens the birth'. So embedded was this figure in the culture of the Middle Kingdom (approximately 2,000 BC) that, it is said, midwives were referred to as 'servants of Heqet'. The frog goddess became a good luck charm for healthy and happy human reproduction, her image appearing on amulets worn by women during childbirth, and commonly represented as a frog's head stuck on the end of a human phallus.

In the centuries that followed, predictably, the Greeks and Romans adopted the Egyptian idea of frogs and toads as symbols of fertility, adding their own cultural appropriations. As well as being symbols of new life, frogs became recognised as creatures that threw caution to the wind sexually, risking everything in the throes of apparent romance and ecstasy. The Greeks and Romans saw in them something of themselves. In this way, anurans became culturally wrapped up with Aphrodite, the Greek goddess of love, beauty, pleasure and (that word again…) fertility. Frogs were now symbols of wild sex, perversions of the cultural ideas that existed thousands of years previously.

A plague of frogs

If only frogs could have embedded themselves deeper into culture at this point, they may have been protected from the smear campaign that was just around the corner. From being creatures of adoration, they went to being creatures of abhorrence. In the sweep of just a few centuries, everything changed, and frogs and toads became something to fear. As related in the Bible (Exodus 8:3), frogs came to be seen as bringers of plagues: 'The Nile will swarm with frogs, which will come up and go into your house and into your bedroom and on your bed, and into the houses of your servants and on your people, and into your ovens and into your kneading bowls.'

Frogs were the second of the ten biblical plagues, the infamous series of calamities inflicted by God upon Egypt to encourage the release of the Israelites from slavery. Although the biblical story is undoubtedly exaggerated, there is certainly a whiff of truth about it – at least when it comes to the frog plague. The fact that amphibian numbers can dramatically rise and fall over years, and that frogs make their own personal exodus from wetlands during periods of rain, can make the sudden appearance of thousands upon thousands of adult frogs a likely event, even in temperate regions.

Above: Gerard Jollain's 1670 interpretation of the Biblical Plague of Frogs.

Can frogs really be considered a plague though? My feeling is that their sudden presence is unlikely to have caused any human suffering. Perhaps frogs were singled out in the story as a way of stamping out belief and mysticism surrounding the Egyptian gods? After all, it was only a few centuries before that the locals were worshipping a frog-headed goddess, and the Ten Commandments decreed that there should be no worship of false idols. Was the plague of frogs a metaphor? Perhaps we may never know, but the age of frogs as creatures of worship was changing, at least in this part of the world.

Ill omens and magical powers

What did the early inhabitants of Britain and Europe make of frogs and toads? The probable answer is that they made dinner from such creatures. Human-gnawed frog bones have been found at a number of archaeological sites across Europe. In the UK, some of these bones date back to 7,000–8,000 BC, long before the French appear to have hit upon this now common delicacy. One thing is clear, however: it wouldn't have been long before these early Europeans came upon the intoxicating nature of toad skin and the profound (if unpredictable and wholly toxic) symptoms its ingestion brings forth. Not surprisingly, toads became synonymous with witchcraft and ill portent, a slur upon all amphibians that persisted right through until Victorian times.

Below: In Medieval Britain, frogs and toads became omens of mystical and occasionally ghoulish portent.

In the early Middle Ages, it seems that these beliefs were taken very seriously. A toad walking across your foot, for instance, immediately brought bad luck, while one appearing in your house was a sign that one of the inhabitants was about to die. Some people apparently spat on toads when they saw them for this reason, while others attempted to scare them away with 'magical' stones left at the entrance to their house. This was a creature to be avoided.

But this was just the start. The suspicious nature with which toads were treated intensified with the proliferation of a strange myth that seems to have arisen in the 12th century. The belief was that toads had a special jewel within their heads, the fabled 'toadstone'. Allegedly, wearers of such stones – which were often kept in a little bag carried on a necklace around their neck – were imbued with magical powers. So embedded in culture did toadstones become, that William Shakespeare gave them a mention in the early 17th century in his play *As You Like It* (Act II, Scene i): 'Which, like the toad, ugly and venomous / Wears yet a

Above: Toadstones can 'cleanse the bowels of filth and excrements' according to *The Book of Secrets*, a thirteenth-century bestseller.

Below: Some toadstones may have been fossilised teeth of *Lepidotes*, an extinct fish group whose fossils are known from Jurassic and Cretaceous clays.

Left: A famous 16th century engraving of a witch feeding her familiars – rats and toads.

precious jewel in his head'. There is, of course, no such stone in a toad's head. Goodness knows what was placed in those little bags.

Yet the fashion for mystical toads was not done just yet. The appetite for the amphibians' magical properties lurched again within the public zeitgeist. For, rather than being cursed, toads began to be seen as a cure. Their secretions became a potential aid to pain and illness, and a possible remedy for disease. Numerous ailments were thought to be aided through the frenzied rubbing of a toad upon afflicted areas. At various times, these amphibians (or rather the secretions they exuded) were used to treat conditions as varied as a sprained wrist, vaginal thrush and even breast cancer – the latter continued in some parts of East Anglia right into the early 20th century. Clearly, these practitioners were onto something, for amphibian secretions have become an important area for medicinal research (see overleaf).

Interestingly, East Anglia was home to another strange myth about toads that persisted into modern times: the power of the so-called toadmen. These unusual characters, referred to as recently as the years following the First World War, apparently possessed the mystical ability to speak to horses. And the magical objects that bestowed them this power? The answer was toad bones, or rather, any single bone from a toad that happened to flow up rather than downstream during a full moon. Needless to say, the toadmen have disappeared – at least for now.

The age of science

From symbols of affection, fertility and sex, to objects of mirth, magic or derision, frogs and toads have found themselves on a rollercoaster journey through human history. Yet in the last two centuries, their relationship with humans has changed again, and a new job has opened up to them. Increasingly, they have become creatures of science: worthy study subjects, compliant, easy to observe and (at least in times past) plentiful. Although many of the experiments on amphibians were undoubtedly brutal (and would fail to pass most ethics committees today) there is no doubt that we have learnt a great deal from them as a result.

Italian investigations

In 1777, Italian scientist Lazzaro Spallanzani investigated the mysterious role of animal semen in reproduction by equipping a selection of fertile male frogs with a pair of special trousers intended to trap their reproductive excretions while they engaged in breeding. The experiment was blissfully simple. When the females released their eggs, Spallanzani observed that the males, clad in their tight-fitting trousers, could not fertilise them. He considered this for a moment before trying another

Below: A modern re-imagining of Spallanzani's contraceptive frog pants. In the original experiments, the garment was wax-coated to ensure no sperm could pass into the surrounding water.

Below right: Scientific illustrations accompanying Spallanzani's frog discoveries, published in 1780.

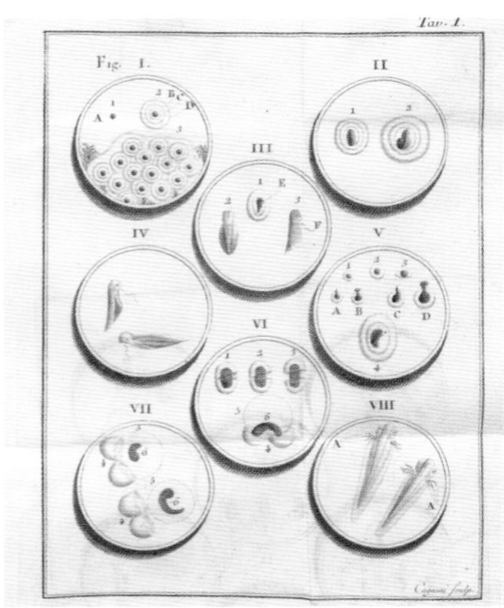

experiment. This time, he poured the contents of the frog pants out onto the eggs, and, hey presto, fertilisation took place. The conclusion? Simple: semen was necessary for female fertilisation. Frogs (and their contraceptive groin-wear) were there right at the beginning of our understanding of how sex works, something for which, I would argue, they deserve our gratitude.

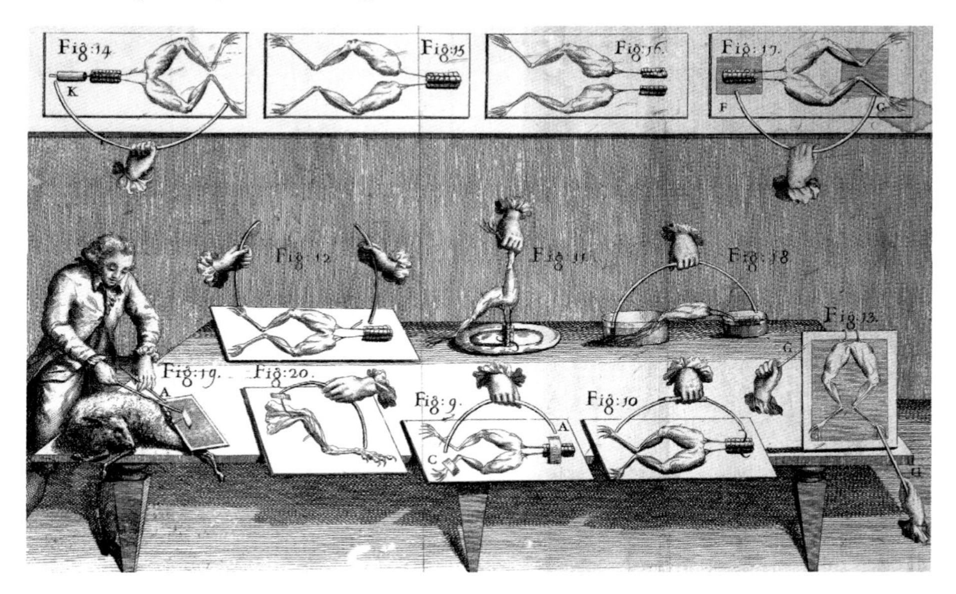

Above: In 1780, Galvani was coming to understand the role that electricity played in animal muscles. A watchful public (that included a young Mary Shelley) delighted in his discoveries.

However, this wasn't the only eureka moment for frog science. Famously, at the University of Bologna in 1780, the assistant of biologist Luigi Galvani accidentally brushed up against a brass hook while touching an experimental dead frog with a steel scalpel. The dead frog's leg twitched in response, which led Galvani (and later experimenters) to investigate the role of electrical impulses and eventually resulted in the discovery of action potentials, the process through which messages travel along nerves. That single lifeless, twitching frog had other far-reaching implications. In its own way, it inspired a young Mary Shelley 40 years later to put to paper the story of a certain monster brought back from the dead through the use of lightning by the scientist Victor Frankenstein.

Above: Hundreds of frogs' legs were once used in medical studies. Though ethically questionable today, body parts like these provided Victorian scientists with crucial insights into a host of biological processes.

Model subjects

Buoyed by early experiments like this, frogs were regularly poked and prodded by anatomists during Victorian times. This is partly because they were so easy to catch and to keep, and partly because they were good models, in an approximate way, for human beings and other vertebrates. With kidneys, a liver, eyes, a brain, a stomach and intestines, amphibians have many of the same internal features of any other bony land animal. Even developing human embryos, in their very early stages, are like those of amphibians. In fact, universities up and down the country used to use as teaching aids whole sets of tadpole embryos carefully chiselled from wax. Today, such antiquated resources have become museum pieces that speak of a former time when amphibian life-history was a way to inspire people into the blossoming science of embryology.

Testing times

Although unethical by our modern standards, the trend for amphibian experimentation continued into recent times. In the 20th century, for instance, frogs were implicit in the discovery of another new research application. They became, of all things, a human pregnancy test. While investigating the role of the pituitary gland in the late 1920s, British biologist Lancelot Hogben noticed something curious about his study animal, the African Clawed Frog, after it was injected with extracts of ox pituitary gland. Inexplicably, the frog began producing eggs. The timing of this discovery was rather fortuitous, because it had recently been revealed that hormones produced by the pituitary, particularly those relating to mammalian pregnancy, were also vented in human urine. Hogben wondered whether concentrated human urine injected into an African Clawed Frog could be a viable way of testing for pregnancy. The answer, as it turned out, was yes. Human urine from pregnant women makes African Clawed Frogs produce eggs, and with gusto.

Within two decades, hospitals all over the world had their own captive African Clawed Frogs in special labs, each regularly injected with doses of female urine to allow doctors to ascertain whether patients were pregnant or not. Between 1940 and 1960, tens of thousands of the frogs were bred for this purpose. But then, quite suddenly, everything changed. A better and more efficient test came along that didn't involve the arduous task of injecting urine into a live frog. The hormones involved in pregnancy were instead detected chemically through a technique still used today – the little white stick. The African Clawed Frogs, some able to live 30 years or more – were now unemployed, and thousands were released into the wild. Sadly, it's possible that some of these frogs may have unwittingly infected wild populations with the disease chytridiomycosis, now known to be the culprit of many amphibian declines worldwide.

Below: Before women could pee on a stick, frogs were the go-to animals for pregnancy testing.

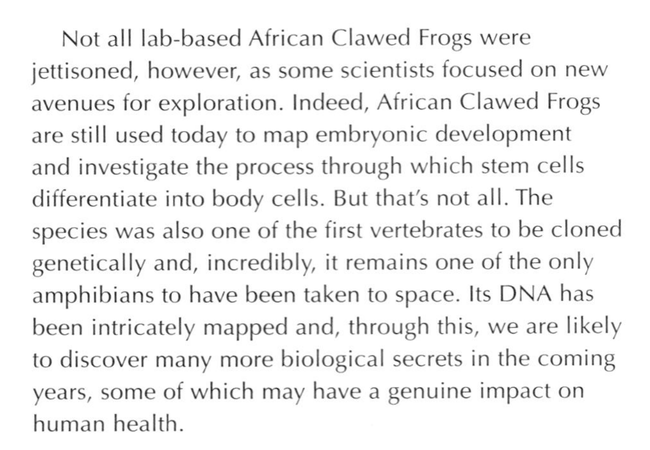

Not all lab-based African Clawed Frogs were jettisoned, however, as some scientists focused on new avenues for exploration. Indeed, African Clawed Frogs are still used today to map embryonic development and investigate the process through which stem cells differentiate into body cells. But that's not all. The species was also one of the first vertebrates to be cloned genetically and, incredibly, it remains one of the only amphibians to have been taken to space. Its DNA has been intricately mapped and, through this, we are likely to discover many more biological secrets in the coming years, some of which may have a genuine impact on human health.

Natural remedies

Although their current global decline is concerning, amphibians are likely to have weathered many pathological storms in the past. In fact, over millions of years they have evolved a suite of skin secretions to kill and eradicate invading pathogens and, interestingly, some of these defences may be of use to us.

Among the amphibian compounds that have been investigated recently by scientists, the most exciting are those produced by frogs and toads to regulate and maintain their skin. The secretions of three closely related Australian tree frogs in the genus *Litoria*, for instance, are of particular interest because of their incredibly effective antibacterial qualities. So strong are these antibacterial secretions that they also appear able to deter some viruses, an immune weapon likened to the holy grail of pathogen defences.

Frog secretions are also of interest to those involved in creating more effective painkillers. The analgesic effects of Amazonian Poison Dart frog secretions have been understood for many years, but alongside them have come some more recent discoveries

Below: Antibiotic secretions produced by the Australian Green Tree Frog (*Litoria caerulea*) are one possible solution to challenge highly resistant strains of superbug that blight many hospitals.

courtesy of Ecuador's Phantasmal Poison-arrow Frog (*Epipedobates tricolor*). In tests on mice, epibatidine – one of the chemicals associated with the secretions of this species – was found to be 200 times more effective than morphine at limiting pain. Efforts are underway to synthesise the chemical artificially, without harming wild frogs, as part of a long-term quest to develop painkillers that are effective yet non-addictive and non-drowsy. Time will tell if frogs, again, will deliver in this quest.

Above: This is one of the most famous dart-frogs, the Reticulated Poison Frog (*Ranitomeya ventrimaculata*). This group of frogs has the potential to revolutionise the effectiveness of painkillers.

And what of anuran poisons? Clearly, the hallucinatory effects of many ingested frog and toad secretions inform us that their chemicals can influence human brain activity. They do this by tweaking a protein vital to the chemical messaging system between cells in the body. The breakdown of this system is at least partly behind such brain conditions as Alzheimer's disease and clinical depression, and there is a small chance that proteins found in amphibian skin could help minimise the breakdown of these pathways.

Below: A key concern of many scientists remains the question of how to explore the chemical secrets of frogs like Ecuador's Phantasmal Poison-arrow Frog without damaging wild populations.

Above: The Australian gastric-brooding frogs, now extinct, provide one example of how much human civilisations have to lose from species extinction.

As fascinating as these medical insights and applications may turn out to be, my personal favourite involves the Australian gastric-brooding frogs, two species in the genus *Rheobatrachus*. Working on laboratory populations of the frogs in the 1990s, scientists managed to determine how they were able to brood their tadpoles in their stomach, an environment laced with digestive enzymes and stomach acids. The key substance involved is called prostaglandin E2, which switches off the body's production of hydrochloric acid, limiting damage to the developing tadpoles. What possible application could this chemical have? Ask the tens of millions of people who suffer from stomach ulcers and heartburn each year. For them, the inner and outer workings of the frogs offer hope of a cure. Yet there is a sting in the tale: these potential amphibian saviours have since become extinct. How many more species we might lose, and what discoveries we might lose with them, is hard to say. It seems clear that the potential contribution frogs and toads can make to human health is another important reason for rallying against the catastrophic declines many are currently facing.

Anurans in popular culture

Left: A fly fisherman: even today, Jeremy Fisher remains a key contributor to the Beatrix Potter estate.

It is time for us to consider two very famous characters, both born in the early 20th century. The first, in 1906, was Jeremy Fisher, a quaint, tidy-looking frog who lived in a 'slippy-sloppy' house on the banks of a pond, dreamt up (and later put to paper) by Beatrix Potter in *The Tale of Mr Jeremy Fisher*. And the second? He was that other famous amphibian, Mr Toad, the rich, motorcar-loving village squire who inhabited the world of Kenneth Grahame's *The Wind in the Willows* (1908). Together, these charismatic characters plundered the literary landscape. Within decades, Fisher's face was on trinkets, cushions, plush porcelain dolls and decorative plates, some still found in antique shops all over the country. Similarly, Mr Toad and his friends became veteran characters in the British children's literary scene, with their story amassing a staggering 31 reprints in just 20 years.

Thanks to Grahame and Potter, the era of amphibians being considered as mean, monstrous, plague-associated creatures was over. Frogs and toads were now re-created as rich characters, easy to exploit for laughs or honest charm. They could be fun (Kermit the Frog), they could be rousing ('We All Stand Together', also known as the

Below: The illustrious (yet foppish) Mr Toad has had many cartoon incarnations. As anatomically correct as Paul Bransom's version is, in later versions, artists decided clothes would be more appropriate.

Frogs: the first cartoon characters?

Above and left: Frogs in the Chōjū-jinbutsu-giga are depicted alongside rabbits and playful monkeys, wrestling and playing. In one climactic scene, a frog is seen praying to Buddha.

In the mid-12th century, around the same time Europeans were running scared from frogs and toads (see page 92), the amphibians' strange charm was being immortalised in Japan in what is one of the world's earliest comics. This is Chōjū-jinbutsu-giga *(Scrolls of Frolicking Animals)*, a famous set of four picture scrolls belonging to Kōzan-ji temple in Kyoto. In these vivid narratives, frogs are pictured wrestling and praying to Buddha, alongside other anthropomorphised everyday creatures. Some art critics consider Chōjū-jinbutsu-giga, with its comic depictions of frogs, the first true 'manga-style' cartoon. Whether or not this is true is still open to debate, but one thing is clear: the cartoon aesthetic of frogs and toads clearly became attractive to other Japanese artists in the centuries that followed.

Above: Kermit and Miss Piggy from *The Muppet Show* arriving at the Oscars.

'Frog Song'), they could be titular stars of arcade games (*Frogger*) and they could be truly awful cartoon popstars (Crazy Frog). They could be used for advertising too. In recent decades, talking animated frogs have sold such things as educational toys, beer ('Bud'... 'wei'... 'ser'), chocolate bars, sweets, coffee and even (inexplicably) garden pesticides.

So what next for anurans on our screens and in our culture? The good news is that frogs and toads appear to be persisting in this, the internet age. The bad news is that they may be undergoing another shift in how they are perceived in the minds of the public. It seems that, online, the image of the humble frog has been co-opted for a new purpose: to sell hateful ideas.

Pepe the Frog started out in 2005 as a 'mellow, positive-vibed frog' in a blissfully simple online comic called *Boy's Club #1*, which captured the hearts of numerous users of the blossoming social networks of

Above: Frog tat or must-have memorabilia? Frogs continue to work a special kind of magic on modern-day consumers.

the time. The image of Pepe, often dashed with ironic quotations of the user's choice, developed an online celebrity status that grew and grew, almost exponentially. By 2015, Pepe was the most recognisable and widely shared meme on the popular social network Twitter. Online, his withering face became impossible to avoid.

But then, dramatically, it all changed. Less than a year later, Pepe's reputation took an unusual turn when, almost totally without warning or build-up, the subversive

Right: From Heqet to hate: an image of Pepe the Frog, an appropriated symbol often used to identify the USA's controversial alt-right movement.

cartoon frog became appropriated by the alt-right movement in the lead-up to the 2016 US elections. Since then, Pepe has become a front-and-centre symbol of this troubling movement, often used online and in print to signal a user's political leanings. Indeed, so bad have things become that, at the time of writing, Pepe's face is now included on the Anti-Defamation League's database of hate symbols. The fortunes of frogs in popular culture can go down as well as up, it would seem.

Above: Money frogs are a popular part of Chinese folklore. It is thought that, pointed at the door from the corner of a room, a money frog can bestow riches upon its owner.

Below: A wooden frog güiro. The stick can be stroked along the frog's back to make a soothing rhythm.

A Future for Frogs and Toads

This chapter offers advice on learning more about amphibians and getting involved in some conservation activities, including pond building, pond surveying and citizen science projects that are helping us limit national frog and toad declines. With your help, the nation's frogs and toads can thrive, so here's what you need to know.

What next for frogs and toads?

Opposite: It's not just for frogs and toads! The modern-day garden pond can be a life-giving oasis to a host of species.

Hopefully, this book has revealed to you how interesting and vital frogs and toads are to the wider world. Not only are they ecosystem dynamos and crucial rungs in the energetics of life, they have a long and rich evolutionary, cultural and scientific history. They are full of charisma, a quality we humans have found attractive, alluring and charming (on the whole) for centuries. Yet, despite this, frogs and toads find themselves in a troubling situation. At the present time, one-third of all amphibian species on Earth are threatened with extinction. We would be foolish to consider them safe in our ponds, lakes and rivers forever. Surely, therefore, now is the time to act.

Below: In 30 years, we've lost two-thirds of our Common Toads. The right kinds of new ponds and a better understanding of their migrations may help to curb further declines.

Doing your bit

There are several things you can do to positively influence the number of frogs and toads that exist across the UK. On the pages that follow, you'll find some ideas to get you started.

Inspire younger generations Anurans and their tadpoles are incredibly accessible creatures. Simply showing young people frogs and toads as part of a community event, or, better still, undertaking some impromptu pond-dipping, is a great way to demonstrate the value of amphibians and the habitats on which they depend.

Record sightings Sightings of frogs and toads matter, always. In fact, recording wildlife observations has become an incredibly important conservation tool in recent years. By plotting recorded observations, both locally and nationally, scientists can chart populations rising and falling, and can step in to act before it is too late. The national wildlife charity Froglife has a free mobile app called Dragon Finder to help you identify and report any

Below: Create some memories! Pond-dipping is an exciting and deeply rewarding activity for both adults and children.

Above: There is no more significant single thing you can do for local nature than create a pond.

amphibians and reptiles you see. Alternatively, if you want to get involved in surveying amphibians more formally, you can become part of the National Amphibian and Reptile Recording Scheme (NARRS), which is run by the Amphibian and Reptile Conservation Trust (see page 125).

Dig a new pond Perhaps the single biggest difference you can make to encourage frogs and toads locally is to dig a pond. Even a tiny pond – little more than a planted-up sink buried in a flower bed – can be an important stop-off for frogs and toads in summer. And ponds aren't just for frogs and toads, of course. They benefit dragonflies, water beetles, caddisflies and thousands of other invertebrates, and they provide vital watering holes and stop-off points for a host of neighbourhood mammals and birds. Crucially, they are places for humans too. Many people find recreational solace beside water, meeting there for picnics and to study the lives of countless other creatures. In this way, ponds are for all of us. The wildlife charity Froglife has an excellent booklet called *Just Add Water* to help get you started, downloadable free online (www.froglife.org).

Join the dots! If you are thinking about digging a pond, consider locating it between two existing frog or toad populations. Even a simple hole in the ground (with a pond liner, if needed) can become a vital stepping stone for local populations, allowing them to intermingle and mix up their genes so that they become more resilient to local habitat changes. The same is true for many other species. Within days of digging and filling a pond, a menagerie of water beetles, damselflies and dragonflies will arrive, eager to colonise the newly available real estate. Networks of ponds really do matter.

Revive an old pond It can be immensely satisfying to revive an old pond that has become filled with silt and old leaves. Smaller ponds can be emptied out with a bucket, swept clean (taking care not to puncture the liner if there is one), and then refilled with water and replanted with natives. The best time of year to revive a pond is in early autumn, after any tadpoles have left as metamorphs and before adult frogs (normally male) come back to spend winter on the bottom. Yes, it is smelly work, but the rewards in spring are well worth it! For tips on improving an existing pond, see page 113.

Below: The only equipment needed to dig a small pond is a spade, a liner and a little bit of elbow grease.

Above: Working together, volunteer amphibian surveyors can help scientists better understand the factors behind widespread frog and toad declines.

Help a toad across the road The national wildlife charity Froglife (www.froglife.org) maintains a database of more than 500 designated crossing points at which some local authorities install special vehicle warning signs during the spring months. At more than a hundred of these sites, small networks of volunteers maintain records of migrating Common Toads and their numbers as part of the citizen science project, Toads on Roads. If you have a few spare nights in early spring, then offering to become part of a toad patrol can be an immensely rewarding way of helping your local amphibians. Not only will you be able to get up close and personal with toads, but the data you record can help scientists understand about recent toad declines.

Survey fresh waters One of the great thrills of pond-dipping is that no one – not even the experts – can predict what they might find. Armed with your net, you might discover nimble fish-like newt larvae, Water Stick Insects, Water Spiders or rare and previously unrecorded leeches not known locally. All such discoveries and observations

Above: Biohazard precautions include giving your boots a good wash after every visit to a pond.

have conservation significance, and there are online resources to help you identify correctly what you have discovered. One way to help out is by participating in the Freshwater Habitats Trust's PondNet project, which brings together networks of pond-dippers to record and monitor widespread species. Your contributions could help secure the future of a freshwater pond and the populations of frogs and toads it might support.

Limit the spread of invasives As mentioned previously (see page 37), the movement of non-native species (and their diseases) is a serious threat to native wildlife, including frogs and toads. Chytridiomycosis and ranaviruses are diseases whose spread could be limited if we think more smartly about how we manage ponds. Although it may be tempting, try to limit movement of frogspawn from pond to pond, and clean nets and wellies thoroughly after pond-dipping. Also, try to avoid buying non-native pond plants – there's a handy list of native pond plants on the Froglife website (see page 126).

Improving ponds

Below are some tips for getting the most from your pond, particularly when it comes to nurturing the tadpoles that may be developing there.

Shallow south-facing edges Tadpoles of both frogs and toads like warmth. Not only do warmer temperatures boost their metabolism, they also encourage the growth of the algae upon which they feed. For this reason, tadpoles may appear to almost bask in the sunny parts of the pond in late spring while they rasp their horny beak-like mouths over the algae beneath. You can create sunlit shallow areas in your pond with large flat rocks, or by intentionally digging a shelved lip around the edge. Forming a mosaic of different layered shelves works particularly well – you'll notice tadpoles dashing into the cracks between the rocks when predators come too close.

Bushy areas with dense branches Densely shaded areas near the pond can be important refuges for metamorphs waiting for the summer rains that see them make their first forays into the wider neighbourhood. A little like an airport departure lounge, hundreds sometimes gather there, silently waiting for suitable

Below: A pond is never just a pond. All of them offer numerous ecological niches by way of the shoreline, the bottom or the surrounding emergent vegetation, for instance.

Above: Take extra care when considering pond plants – there are many invasive species that you will need to avoid. Native pondweeds include Hornwort (left) and Curled Pondweed (right).

conditions. If your pond edge is decorated with paving slabs, beware that these can get very hot in the sun, to the extent that they effectively cook young froglets alive as they emerge from the water. To reduce the chance of this happening, plant shady foliage, install a solar-powered fountain to keep the pavers damp, or roll out a strip of turf over the pavers before the froglets climb out.

Good access Some pond designs make it difficult for amphibians to get out of the water. This can be a problem for metamorphs particularly, which have limited energy reserves and so are running on borrowed time. A simple 'stairway' leading out of the pond, made of bricks or decorative rocks, can be a very helpful addition.

Pond plants Though their impact on oxygen levels may not be all they are hyped up to be, oxygenating plants in the water do provide important niches for invertebrates to colonise and tadpoles to hide among. You can think of them as offering reef-like diversity for your pond

Left: Solar-powered fountains can add a decorative touch to some ponds. Just remember that they will increase the rate of evaporation from the pond.

inhabitants. Native species worth your interest are Curled Pondweed (*Potamogeton crispus*) and Rigid Hornwort (*Ceratophyllum demersum*). Stock your cuttings from trusted neighbours. The more local, the better.

Use freshwater Increasingly, pollution in ponds is known to affect freshwaters across the country. You can help by ensuring your pond is topped up with rainwater (from out of a waterbutt) rather than from out of the tap. Because tapwater contains nutrients, using rainwater can restrict the rampant growth of problem plants – including blanketweeds and duckweeds.

Reconsider fish! Although many pond owners love their ornamental fish, these are often predators of spawn, tadpoles and aquatic invertebrates. Introducing fish therefore limits the biodiversity in wildlife ponds, and many frogs and toads may choose to spawn elsewhere. For this reason, some gardeners choose to dig two ponds – one for fish and one for wildlife!

Watching frogs and toads

Above: Frogs have such variation within populations that many local individuals can be told by sight. Naming is optional.

If, by this point in the book, you feel your heart stirring for frogs and toads, there are some simple strategies you can employ to see them regularly. Below are my choice tips for finding anurans in your local area.

Think prey Remember that amphibians are attracted to damp places where invertebrates gather. At almost any time of year, your most likely chance of seeing amphibians is by gently lifting logs (particularly) or bits of bark or other detritus that lie within 25m (27yd) of a pond.

Think sun Like reptiles, frogs sometimes like to bask and can occasionally be observed warming themselves up on sunny mornings. In late spring, if you get up early and look on the north-western side of your local pond, you might have a lucky sighting.

Think spring To see frogs and toads at their best, visit any small or medium-sized pond (for frogs), or medium-sized to large pond (for toads), in late January or February (in the south), or March or even April (in the north and east). Amphibians make their migrations on mild nights, ideally following a bit of afternoon rain. Generally speaking, the milder the temperatures, the better. Websites like *Nature's Calendar* (naturescalendar.woodlandtrust.org.uk) provide information on spawning around the country. In some years, it can take place as early as December in south-western parts of the UK.

Don't forget autumn Another period of amphibian activity occurs in late September. During spells of warm rain, younger frogs (particularly) make movements through long grass away from ponds. If you happen to be walking near a ditch or in a local park at this time of year, or even near a water feature of any size, you might come across them scurrying through nearby patches of grass.

Put in a pond Perhaps the best way to see frogs and toads is to bring them to you by digging a pond. Even in urban areas, frogs are quite quick to colonise new ponds – incredibly, the first visits can occur within a matter of days.

Visit at night Taking a torch to a local pond at night can be an utterly riveting experience, exposing all sorts of pond creatures you may never before have realised were there, including water beetles, newts and even bats hunting at the water's surface. Amphibian lovers call this torching, and it is an important technique for surveying amphibians in the wild.

Below: An example of one of many hedgehog holes appearing across the country. Frogs and toads undoubtedly use these too.

Pond-dipping

Dipping a net into a pond in spring or summer is a fantastic way to get a feel for the creatures that live within, including tadpoles. If it is done carefully and considerately, the animals will remain unstressed, meaning you can take a few moments to see their wild behaviour up close. Although you can buy pond-dipping nets and trays fairly easily, some simple kitchen items do the job just as ably.

Kitchen sieve Waving a clean kitchen sieve beneath the surface of your pond can reveal a great deal. Many of the most interesting pond creatures are also the tiniest, including water fleas and fly larvae, so the smaller the holes in your sieve, the better.

White ice cream tub A water-filled white ice cream tub is ideal for holding the netted creatures while you observe them. Fill the container with a few centimetres of pond water before emptying the contents of your sieve into it. You should be able to see your pond creatures swirling around, exposing themselves to your watchful gaze. This is a real moment of magic. Be careful not to leave the animals in the tub for more than a few minutes, particularly on hot days, and don't put obvious predators in trays with obvious prey – dragonfly nymphs, for instance, will make short work of captive tadpoles if left for too long!

White plastic spoon A children's medicine spoon doubles up perfectly as a device for temporarily scooping up animals like tadpoles for a closer look with a magnifying glass. Remember to wash it carefully afterwards.

Change of clothes Spare clothes are vital for those inevitable 'Whoops, I fell in!' moments, particularly when children are involved. (Remember also that ponds can be dangerous: it goes without saying that children shouldn't play in or near ponds without supervision.)

Wash-down It's always good practice to wash your hands thoroughly after pond-dipping, using antibacterial soap. Also, remember to clean your kit thoroughly before using it in another pond – the movement of invasive species and their diseases is a big threat to amphibians.

Below: Pond-dipping helps pupils learn about a variety of curriculum topics including food chains, classification, adaptations, local habitats and ecosystems.

The national picture

Thanks to the work of naturalists, the situation with the UK's declining amphibians is becoming clearer. Although Common Frogs appear to have adapted well to our modern taste for garden water features, our toads have not. Common Toads, in particular, are sticklers for tradition, and it may be that this dogged faithfulness to old breeding sites is partly behind their apparent decline in many parts of the country. And when toads do attempt to discover new habitats, modern infrastructure (particularly roads) can create impassable boundaries that restrict local population growth. Every man is an island, and the same is true of many Common Toads and their ponds. Essentially, if a pond dies and there is no connection to another one, there is no escape. This is essentially what happened to our Pool Frogs and Natterjack Toads decades ago, but now writ large.

Is there a solution to the problematic situation in which many toads find themselves? The answer is... possibly. Clearly, amphibian habitats need to be linked

Below: Wildlife connectivity is starting to seep into the public conscious. Here, a virtual reality experience created by the charity Froglife takes users on a simulated journey underneath a road.

up, so that amphibians can move more freely between breeding grounds and populations can be buffered from the threat of local extinction. The question is, how can this be done in a landscape where the needs of humans are rated so much higher than the needs of amphibians? Politically, can the aspirations of conservationists win through to benefit the lives of the animals that they represent? The encouraging news is that this may indeed be possible, given time.

Wildlife corridors

In recent years, a number of conservation organisations have begun collaborating more closely in order to assess ways in which fragmented habitats can be linked to one another. This is being done through the creation of wildlife corridors, specially designated strips of habitat that act like bridges (or sometimes tunnels) between habitats.

The RSPB is one of the UK's wildlife organisations committed to encouraging this concept, and indeed it offers support and advice in creating nature byways and highways through neighbourhoods that promote the

Below: Large wildlife bridges like this one allow amphibian populations on either side of the road to continue mixing genetically.

Above: Green bridges like this on in East London offer wildlife a crucial pathway through urban areas.

connection of important habitats separated by urban infrastructure. Other charities that have taken on the approach include the British Hedgehog Preservation Society and the People's Trust for Endangered Species, which together promote Hedgehog Street (www.hedgehogstreet.org), a campaign that helps local communities link up garden Hedgehog habitats, thereby bolstering local populations. As an added bonus, campaigns like these also help our frogs and toads persist.

On a wider level, the charity Froglife (along with European partners) has been undertaking research into the effectiveness of toad tunnels, which are a potentially favourable measure to connect sites that are dissected by roads. Their results will be of interest to many. Indeed, toad tunnels can offer valuable connections for other British wildlife, including voles, Otters, Hedgehogs, reptiles and locally threatened invertebrates.

Above: A forgotten pastime? Our job as modern wildlife lovers is to pass on our love of nature to those generations that will follow.

Opposite: Saving frogs and toads depends on reconnections, both for the amphibians themselves but, increasingly, for us too.

If I have one dream for the future of frogs and toads in the UK, it is that we consider the places where they are not present as much as the places they are. There is no reason why frogs and toads could not be resident in even the most urban parks in Britain and Ireland, where invertebrate life can abound if we wish it so. Through the use of carefully planned wildlife corridors and tunnels, we can encourage frogs and toads to move back to these areas, no matter how built up they may be. This applies as much to threatened Pool Frogs and Natterjack Toads as to their more widespread cousins, the Common Frog and Common Toad. Perhaps the future of these charismatic species will be closely allied to our own, with shared spaces and places to grow, and a mutual thirst for life quenched. Perhaps one day, it won't be us that saves frogs and toads, but rather they that save us from ourselves. I certainly believe so. I hope you, upon reading this book from start to finish, have come to same conclusion.

Glossary

Aestivation A prolonged period of dormancy activated during hot or dry spells of weather.

Amplexus The mating position used by most frogs and toads, whereby males clasp females from behind, often gripping behind the female's front legs.

Anuran An amphibian representative of the taxonomic order of creatures that includes all frogs and toads.

Apoptosis The coordinated cell death that occurs in the tails of tadpoles prior to the completion of metamorphosis.

Chytridiomycosis An infectious disease in amphibians caused by the chytrid fungus *Batrachochytrium dendrobatidis*.

Cloaca The waste vent in most non-mammalian vertebrates through which excretory products and eggs or sperm are released.

Dorsal The upper side or back of an animal, plant or organ.

Froglet See 'metamorph'.

Herpetologist A scientist who studies amphibians and/or reptiles.

Invertebrate An animal of any type that lacks the distinctive nerve cord running from head to tail.

Metamorph The descriptive term for an amphibian immediately after metamorphosis; also called a froglet or toadlet.

Metamorphosis The distinct transformation from an immature life stage to an adult (usually sexual) form.

Nuptial pad The swelling on the inner forefingers of male frogs and toads that allows for extra grip during amplexus.

Parasite An organism that lives upon or within another organism, from which it derives its nutrients at a cost to the host.

Parotoid gland An external skin gland found upon toads (and some frogs) that produces unpleasant and toxic chemicals.

Permeable A material or membrane through which liquids or gases can pass.

Ranavirus A genus of viruses in the family Iridoviridae, capable of infecting amphibians and reptiles.

Toadlet See 'metamorph'.

Unken reflex The defensive 'play dead' posture adopted by some amphibians when stressed.

Ventral The lower side of an animal, plant or organ.

Vertebrate An organism with a central nerve cord that runs from head to tail, often surrounded by bones (vertebrae) that make up a spinal column.

Further Reading and Resources

Books

An impressive array of books covering British amphibians in all their glory is available, with each providing plenty of information about their behaviour and ecology, as well as guidance on their conservation. Below is a handful of my favourites.

Beebee, T. 2013. *Amphibians and Reptiles*. Naturalists' Handbooks 31. Pelagic Publishing, Exeter.
In this go-to guide, Trevor Beebee covers the biology, ecology, conservation and identification of British amphibians and reptiles, and provides keys for the identification of amphibian adults, their eggs, larvae and metamorphs. This is a useful edition to the bookshelf of any wannabe 'herper'.

Beebee, T. & Griffiths, R. 2000. *Amphibians and Reptiles*. New Naturalist Library 87. HarperCollins, London.
Respected herpetologists Trevor Beebee and Richard Griffiths bring together a wealth of new and fascinating information on the British amphibians and reptiles in this important addition to the long-running New Naturalist series.

Crump, M. 2015. *Eye of Newt and Toe of Frog, Adder's Fork and Lizard's Leg: The Lore and Mythology of Amphibians and Reptiles*. University of Chicago Press, Chicago and London.
If you enjoyed the chapter 'Frogs and Toads in Culture', you might also appreciate this offering from science writer Marty Crump. It's a fascinating journey across cultures, exploring the complicated relationship humans have had down the years with frogs, toads, other amphibians and reptiles.

Inns, H. 2011. *Britain's Reptiles and Amphibians*. Princeton University Press, Princeton.
A superbly illustrated guide (in the WILDGuides series) ideal for those eager to learn more about species identification, behaviour and, vitally, the successful conservation of British amphibians and reptiles. Aimed at amateurs and professionals alike, this colourful guide is one I return to again and again.

Minting, P. & McInerny, C. 2016. *Amphibians and Reptiles of Scotland*. Glasgow Natural History Society, Glasgow.
Produced by the Amphibian and Reptile Conservation Trust, this book offers up-to-date information about frogs and toads in Scotland. It is available as a free download from Glasgow Natural History Society (www.glasgownaturalhistory.org.uk).

Speybroeck, J., Beukema, W., Bok, B., Van Der Voort, J. & Velikov, I. 2016. *Field Guide to the Amphibians and Reptiles of Britain and Europe*. Bloomsbury, London.
The UK's amphibians can be better understood when considered alongside the wealth of varied and interesting species that exist in Europe. This guide includes information on ecology, distribution and identification for every European species. There's even a handy tick list.

Conservation groups

Amphibian and Reptile Conservation Trust (ARC)
www.arc-trust.org
The ARC is an independent non-governmental organisation committed to the conservation of amphibians and reptiles. In addition to managing a suite of important sites for rare species, the trust has been a key player in reintroducing

once lost native species like the Pool Frog back to the wild.

Amphibian and Reptile Groups of the UK (ARG UK)
www.arguk.org
An umbrella organisation supporting local amphibian and reptile groups in many parts of the country that offer opportunities to engage in training and local monitoring of frogs and toads, as well as amphibians and reptiles more generally. There is also a national conference each year and a website with plenty of information. ARG UK is a one-stop shop for those wanting to engage in the local conservation of frogs and toads.

Freshwater Habitats Trust
www.freshwaterhabitats.org.uk
The Freshwater Habitats Trust's aim is to protect freshwater life for everyone to enjoy. Its website is a mine of important information about ponds and pond building. As well as running citizen science projects such as PondNet to monitor water pollution and assess the status of widespread species, the trust also puts together many easy-to-digest best-practice guides. One of its most significant projects is the Million Pond Project, which aims to restore Britain's wetlands after a century of pond losses.

Froglife
www.froglife.org
Froglife runs numerous projects across the country, many creating new ponds and renovating existing ones. As well as providing practical opportunities to engage with wildlife in London, Peterborough and Glasgow, the charity also runs Toads on Roads, a national database of amphibian crossing sites throughout the UK.

Garden Health Initiative
www.gardenwildlifehealth.org
This long-running project encourages the public to observe incidents of unusual mortality in garden species, including frogs and toads. Since its inception, it has helped improve understanding of a host of complex diseases, including, in frogs, the condition often referred to as red-leg, which afflicts local populations in summer months.

Royal Society for the Protection of Birds (RSPB)
www.rspb.org.uk
The RSPB works for a healthy environment rich in birds and other wildlife. As well as managing a fleet of top-class nature reserves in which threatened amphibians flourish, the society has also become an important player in influencing government on behalf of many of the creatures that make their home in the UK. Its website is a superb place to find guidance on pond building and making gardens more nature-friendly.

The Wildlife Trusts
www.wildlifetrusts.org
Like Amphibian and Reptile Groups of the UK, local Wildlife Trusts are an important resource providing information on amphibians and where to find them. There are 47 local Wildlife Trusts across the UK, including the Isle of Man and Alderney. Most Wildlife Trusts offer training courses and also run family events and local projects, including pond building and pond maintenance.

Acknowledgements

Being given permission to write about frogs and toads has been a great privilege. Enormous thanks, therefore, must go to Julie Bailey at Bloomsbury for the commission, my editor Alice Ward for steering me so expertly through its production and Jim Martin for his continued support and kindness throughout.

I am indebted to Professor Roger Downie of Glasgow University for kindly looking through the text of this book in its final stage. Roger provided a host of informative and enlightening suggestions, all of which improved my understanding of these fascinating creatures no end. Big thanks also to Susi Bailey, whose insightful comments and various edits on the text improved the look and feel of the book hugely.

I am enormously grateful to a number of other people who have encouraged my love of UK amphibians, professionally and non-professionally, either in person or through their books and guides. Their words over the years have influenced this book very much. These people include Rob Oldham, Arnie Cooke, Tony Gent, Roy Bradley (whose pond has always been an inspiration!), Jim Foster, Tom Langton, Daniel Piec, Ruth Carey, Brian Laney, John Wilkinson, Sam Goodlet, Howard Inns and Trevor Beebee. Thanks especially also go to Kathy Wormald and all at Froglife for their continued support. I am proud to be the co-patron of this excellent wildlife charity.

Lastly, huge thanks go to those who encouraged my love of frogs and toads from a very early age, and let me simply get out there and explore these wonderful animals for myself… as long as I was back in time for tea! I am referring here, of course, to my mum and dad – this one's for you both. x

Image credits

Bloomsbury Publishing would like to thank the following for providing photographs and for permission to reproduce copyright material. While every effort has been made to trace and acknowledge all copyright holders, we would like to apologise for any errors or omissions and invite readers to inform us so that corrections can be made in any future editions of the book.

Key t = top; l = left; r= right; tl = top left; tcl = top centre left; tc = top centre; tcr = top centre right; tr = top right; cl = centre left; c = centre; cr = centre right; b = bottom; bl = bottom left; bcl = bottom centre left; bc = bottom centre; bcr = bottom centre right; br = bottom right

AL = Alamy; FL= FLPA; G = Getty Images; NPL = Nature Picture Library; RS = RSPB Images; SS = Shutterstock, iStock = iS

Front cover t Sue Kennedy/RS, b Mark Bridger/G, **back cover** t Mike Powles/G, b Ben Andrews; **1** AbiWarner/G; **3** SS; **4** Malcolm Schuyl/FL; **5** Simon J Beer/iS; **6** Mike Read/RS; **7** National Geographic Image Collection/Al; **8** Premaphotos/Al; **9** Daniel Heuclin/NPL; **10** SS; **11** Jelger Helder, Buiten-beeld/FL; **12** Jelger Helder, Buiten-beeld/FL; **13** SS; **14** l SS, r Stephen Dalton/NPL; **15** SS; **16** Michael Dietrich/FL; **17** Dale Sutton/2020VISION/NPL; **18** golfer2015/iS; **19** t Jane Burton/NPL, Alamy Stock photo; **20** Jack Perks/RS; **21** b Rob_Ellis/iS, t Willem Kolvoort/NPL; **22** t Imagebroker/FL, Ben Hall/NPL; **23** SS; **24** Alex Hyde/NPL; **25** SS; **26** IvonneW/iS; **27** t Norfolk Wildlife Trust, b Wild Wonders of Europe/Widstrand/NPL; **28** Klaas van Haeringden/Minden Pictures/FL; **29** Pat Tuson/NPL; **30** George McCarthy/RS; **31** Simon Litten/FL; **33** Dave Pressland/FL; **34** AlasdairJames/G, SS; **35** Jelger Helder, Buiten-beeld/FL; **36** Â© Biosphoto/FL, Denis Palanque/FL; **37** Alex Huizinga/FL; **38** SS; **39** John Cancalosi/NPL; **40** SS; **41** SS; **42** SS; **43** SS; **44** SS; **45** George McCarthy/NPL; **46** Nick Upton/NPL; **47** t blickwinkel/AL, b SS; **48** Inaki Relanzon/NPL; **49** Paul Hobson/NPL; **50** SS; **51** Chris Hennessy/AL; **52** John Cancalosi/NPL; **53** Jojo/Wikipedia; **54** t SS, b Michael Hutchinson/NPL; **55** SS; **56** t Dave Pressland/FL, b Erica Olsen/FL; **57** t Jelger Herder/Buiten-beeld/Minden Pictures/G, b Willem Kolvoort/NPL; **58** Cyril Ruoso/Minden Pictures/FL; **59** l SS, r Jose Luis Gomez de Francisco/NPL; **60** SS; **61** Cyril Ruoso/NPL; **62** SS; **63** t Dr David M. Phillips/G, b Ashley Cooper/NPL; **64** Stephen Dalton/G; **65** t Jose Luis Gomez de Fancisco/NPL; b emer1940/iS; **66** SS; **67** Stephane Vitzthum/FL; **68** t blickwinkel/AL, b SS; **69** t Maximillian Weinsieri/AL, b philgood/G; **70** tl Naturfoto Honal/G, tr Valter Jacinto/G, bl Wild Wonders of Europe/Widstrand/NPL, br Emanuele Biggi/NPL; **71** Lizzie Harper; **72** Dave Pressland/FL; **73** SS; **74** Steve Trewhella/FL; **75** t Richard Bowler/RS, b Duncan Mcewen/NPL; **76** SS; **77** Oregan State University/Wikipedia; **78** SS; **79** Anton Sorokin/AL; **80** Nick Hamilton Photographic/AL; **81** Xinhua/AL; **82** t Breaking New Ground, b Dave Bevan/NPL; **83** t Rachel Husband/AL, b James Lowen/AL; **84** t Jules Howard, b Gary K Smith/NPL; **85** Jules Howard; **86** Heritage Images/G; **87** Rogers Fund, 1918; **88** Olaf Tausch/Wikipedia; **89** Andrew®/Wikipedia; **90** Gerard Jollain/Wikipedia; **91** Unknown/Wikipedia; **92** Stefano Bianchetti/G, The Natural History Museum/AL; **93** Unknown/Wikipedia; **94** l © Maquettree Studios, Maquetas lda. Model permanently exhibited at Museu da Ciência da Universidade de Coimbra, r Spallanzani, Lazzaro/Wikipedia; **95** Welcome Collection gallery/Wikipedia; **96** Leon Neal/Staff/G; **97** National Museum of Health and Medicine/Science Photo Library; **98** SS; **99** SS; **100** © NHPA/Photoshot; **101** t Archivah/AL, b Paul Bransom/Wikipedia; **102** Toba Sojo/Wikipedia; **103** SS; **104** Josh Edelson/G, ledaphne/iS; **105** SS; **106** Mike Birkhead/G; **107** SS; **108** Adrian Sherratt/AL; **109** SS; **110** Alphotographic/iS; **111** ajsissues/AL; **112** bitenka/iS; **113** SS; **114** SS; **115** whfnew; **116** Paul Hobson/G; **117** Lisa Edie/AL; **118** SolStock/iS; **119** Jules Howard; **120** SS; **121** Photofusion/G; **122** Gary K Smith/FL; **123** Jelger Herder, Buiten-beeld/FL.

Index